Military Strategy in the 21st Century

Military Strategy in the 21st Century explores military strategy and the new challenges facing Western democracies in the twenty-first century, including strategy in cyber operations and peacekeeping, challenges for civil-military relations, and the strategic choices of great powers and small states.

The volume contributes to a better understanding of military strategy in the twenty-first century, through exploring strategy from three perspectives: first, the study of strategy, and how our understanding of strategy has changed over time; second, new areas for strategic theory, such as peacekeeping and cyberspace; and third, the makers of strategy, and why states choose sub-optimal strategies.

With the increasing number of threats challenging strategy makers, such as great power rivalry, terrorism, intrastate wars, and transnational criminal organisations, *Military Strategy in the 21st Century* will be of great value to scholars of IR, Security Studies, Strategic Studies, and War Studies as well as policymakers and practitioners working with military strategy in particular and international security and war in general. The chapters were originally published as a special issue of the *Journal of Strategic Studies*.

Kersti Larsdotter is Associate Professor of War Studies at the Swedish Defence University. Her research includes the dynamics, nature, and conduct of war, specifically civil wars and different forms of military interventions. She has published in journals such as the *Journal of Strategic Studies*, *Small Wars & Insurgencies*, and *Parameters*.

Military Strategy in the 21st Century

Edited by
Kersti Larsdotter

Routledge
Taylor & Francis Group

LONDON AND NEW YORK

First published 2020
by Routledge
2 Park Square, Milton Park, Abingdon, Oxon, OX14 4RN

and by Routledge
52 Vanderbilt Avenue, New York, NY 10017

Routledge is an imprint of the Taylor & Francis Group, an informa business

First issued in paperback 2021

British Library Cataloguing in Publication Data
A catalogue record for this book is available from the British Library

ISBN 13: 978-0-367-44153-1 (hbk)
ISBN 13: 978-1-03-208405-3 (pbk)

Typeset in Myriad Pro
by Newgen Publishing UK

Publisher's Note
The publisher accepts responsibility for any inconsistencies that may have arisen during the conversion of this book from journal articles to book chapters, namely the inclusion of journal terminology.

Disclaimer
Every effort has been made to contact copyright holders for their permission to reprint material in this book. The publishers would be grateful to hear from any copyright holder who is not here acknowledged and will undertake to rectify any errors or omissions in future editions of this book.

Contents

Citation Information

The chapters in this book were originally published in the *Journal of Strategic Studies*, volume 42, issue 2 (April 2019). When citing this material, please use the original page numbering for each article, as follows:

Chapter 1
Military strategy in the 21st century
Kersti Larsdotter
Journal of Strategic Studies, volume 42, issue 2 (April 2019), pp. 155–170

Chapter 2
Strategy in theory; strategy in practice
Hew Strachan
Journal of Strategic Studies, volume 42, issue 2 (April 2019), pp. 171–190

Chapter 3
Military strategy and peacekeeping: An unholy alliance?
Kersti Larsdotter
Journal of Strategic Studies, volume 42, issue 2 (April 2019), pp. 191–211

Chapter 4
Fancy bears and digital trolls: Cyber strategy with a Russian twist
Benjamin Jensen, Brandon Valeriano and Ryan Maness
Journal of Strategic Studies, volume 42, issue 2 (April 2019), pp. 212–234

Chapter 5
The political-military dynamic in the conduct of strategy
John Kiszely
Journal of Strategic Studies, volume 42, issue 2 (April 2019), pp. 235–258

Chapter 6

Trigger happy: The foundations of US military interventions
Michael Mayer
Journal of Strategic Studies, volume 42, issue 2 (April 2019), pp. 259–281

Chapter 7

Weak party escalation: An underestimated strategy for small states?
Jan Angstrom and Magnus Petersson
Journal of Strategic Studies, volume 42, issue 2 (April 2019), pp. 282–300

For any permission-related enquiries please visit:
www.tandfonline.com/page/help/permissions

Notes on Contributors

Jan Angstrom is Professor of War Studies at the Swedish Defence University in Stockholm, Sweden. His research interests mainly cover issues related to the use of force. His latest book (with J.J. Widén) is *Contemporary Military Theory: The Dynamics of War* (Routledge, 2015).

Benjamin Jensen holds a dual appointment as Associate Professor at Marine Corps University and as Scholar-in-Residence at American University, School of International Service. He is also a senior nonresident fellow at the Atlantic Council.

John Kiszely served in the British Army for 40 years rising to the rank of lieutenant general. He was Deputy Commander of NATO forces in Bosnia (2001–02) and of the coalition in Iraq (2004–05); he also served three tours of duty in the UK Ministry of Defence, latterly as Assistant Chief of the Defence Staff. His final appointment was that of Director General of the UK Defence Academy. From 2013–17 he was a visiting fellow at the Changing Character of War Centre at Oxford University while writing his book, *Anatomy of a Campaign: The British Fiasco in Norway, 1940* (2017).

Kersti Larsdotter is Associate Professor of War Studies at the Swedish Defence University. Her research includes the dynamics, nature, and conduct of war, specifically civil wars and different forms of military interventions. She has published in journals such as the *Journal of Strategic Studies*, *Small Wars & Insurgencies*, and *Parameters*.

Ryan Maness is Assistant Professor at the Naval Postgraduate School.

Michael Mayer is an independent researcher and author specialising in US security policy and the strategic implications of military technology. His most recent book, *US Missile Defense Strategy*, was published by Lynne Rienner in 2015.

Magnus Petersson is Professor of Modern History and Head of the Center for Transatlantic Studies at the Norwegian Institute for Defence Studies in Oslo, Norway. His research interest is mainly connected to Nordic defence and security policy. His latest book is *The US NATO Debate* (Bloomsbury, 2015).

Hew Strachan has been Professor of International Relations at the University of St Andrews since 2015. He is a Life Fellow of Corpus Christi College, Cambridge, where he taught from 1975 to 1992. In 2016 he was awarded the Pritzker Literature Award for Lifetime Achievement in Military Writing. His recent publications include *The Politics of the British Army* (1997); *The First World War: To Arms* (2001); *The First World War: A New Illustrated History* (2003); and *The Direction of War* (2013).

Brandon Valeriano is the Donald Bren Chair of Armed Politics at Marine Corps University.

Military strategy in the 21st century

Kersti Larsdotter 🄳

ABSTRACT
This special issue explores military strategy in the twenty-first century. The articles scrutinise strategy from three perspectives: the study of strategy, and how our understanding of strategy has changed over time; new areas for strategic theory, i.e., areas where the development of war has made strategy become more important, such as peacekeeping operations and cyberspace; and the makers of strategy, more specifically why states choses suboptimal strategies and how wars in the twenty-first century influence strategy makers.

The development of international security and the conduct of war in the twenty-first century has proven highly problematic for strategy makers. The increasing number of nonstate threats, such as terrorism, intrastate wars, and transnational criminal organisations, the changing norms of intervention, as well as the blurring of lines between war and peace, have challenged strategy, both in theory and practice. Western democracies intervening in intrastate conflicts have received sharp criticism. Not only have they been criticised for pursuing the wrong strategies in contemporary conflicts, such as those in Afghanistan and Iraq, they have also repeatedly been accused of not having a strategy at all, or at least not clearly stated political goals, for the use of military force in these interventions.[1]

The aim of this special issue is to contribute to a better understanding of military strategy and the challenges facing Western democracies in the twenty-first century. We will do that by exploring strategy from three perspectives. The first focuses on the study of strategy, and on how our understanding of strategy has changed over time. The second perspective focuses on new areas for strategic theory, i.e., areas where the development

---

[1]See, for example, Tim Bird and Alex Marshall, *Afghanistan: How the West Lost its Way* (New Haven, CT: Yale University Press 2011); Alastair Finlan, *Contemporary Military Strategy and the Global War on Terror: US and UK Armed Forces in Afghanistan and Iraq 2001–2012* (London: Bloomsbury Academic 2014); Hew Strachan, 'Strategy or Alibi? Obama, McChrystal and the Operational Level of War', *Survival* 52/5 (2010), 157–182.

of war has made military strategy more important, such as peacekeeping operations and cyberspace. Finally, the last perspective focuses on the makers of strategy, more specifically, on why states choses suboptimal strategies, and how wars in the twenty-first century influence strategy makers.

This special issue contains the papers of the conference on 'Military Strategy in the 21st Century' at the Norwegian Defence Command and Staff College in Oslo 13th–14th June 2017. The conference is the fourth since the first Doctrine Conference in Oslo in 2014, the papers from which were published in the *Journal of Strategic Studies* Vol. 39, 2016 Issue 2. The second conference 'Mission Command—Wishful Thinking?' explored historical and contemporary issues of mission command. The papers are part of an edited volume published by the Royal Swedish Academy of War Sciences.

The study of strategy

Strategic Studies is a multidisciplinary and rather young academic discipline, but with roots in a long tradition of the study of strategy in military academies. Although relatively well established today,[2] fundamental questions about the development and identity of the discipline are continuously debated, some of which are particularly pertinent to the development of wars in the twenty-first century.

Although strategy is usually understood as the relationship between ends and means, the nature of this relationship is frequently debated.[3] One of the main questions concerns the relation between the military and political levels. At one end of the scale is Carl von Clausewitz with his understanding of strategy as 'the use of the engagement for the purpose of the war'.[4] This is usually contrasted with an understanding of strategy as the use of war for the purposes of policy, i.e., that battle is not an end in itself but rather a means to an end. The tension between the military and political levels has resulted in the development of different concepts of strategy, such as 'military strategy', focusing on the military level of war, as well as 'grand

---

[2]The number of introductory books in Strategic Studies has, for example, increased over the last 10–15 years. See, for example, John Baylis, James Wirtz, Eliot Cohen and Colin Gray (eds.), *Strategy in the Contemporary World: An Introduction to Strategic Studies* (Oxford: Oxford University Press 2002); Thomas M. Kane and David J. Lonsdale, *Understanding Contemporary Strategy* (London: Routledge 2012); Thomas G. Mahnken and Joseph A. Maiolo (eds.), *Strategic Studies: A Reader* (London: Routledge 2008); Elinor C. Sloan, *Modern Military Strategy: An Introduction* (London: Routledge 2017).
[3]For an overview of several definitions, see John Baylis and James J. Wirtz, 'Introduction', in John Baylis, James Wirtz, Eliot Cohen and Colin Gray (eds.), *Strategy in the Contemporary World: An Introduction to Strategic Studies* (Oxford: Oxford University Press 2002), 1–14, 4.
[4]Carl von Clausewitz, *On War*, translated and edited by Michel Howard and Peter Paret (Princeton, NJ: Princeton University Press 1976), 177.

strategy' and 'national strategy', focusing on the political level, including other means than military.

Another important question is whether strategy should be understood as the 'instrumental link between military means and political ends' or as the 'process by which military objectives and force levels are set'.[5] Both understandings have merit. Understanding strategy as the former makes us focus on *how* military force can be used to achieve political or military objectives. Richard K. Betts, for example, defines strategy as 'a plan for using military means to achieve political ends'.[6] Other scholars understand strategy more in terms of a 'theory of victory' or 'theory of success', emphasising the causal mechanism between ends and means.[7] By understanding strategy as a link between ends and means, the discussion about levels mentioned above becomes less important. It is possible to include several levels of analysis, 'from maneuvers of units in specific engagements through larger campaigns, whole wars, grand strategies, and foreign policies', Betts argues, as long as focus is on 'the *linkages* in the hierarchy of policy, strategy and operations, where the logic at each level is supposed to govern the one below and serve the one above'.[8]

Instead, by understanding strategy as a process, the focus is turned to the actors conducting strategy and the relationship between them. Basil Liddell Hart, for example, defines strategy as 'the art of distributing and applying military means to fulfill the ends of policy'.[9] Who the strategy makers are, how they develop strategies and what influence their decision-making processes, are all significant questions. Here, the levels discussed above become more important. If one understands strategy as the use of the battle for the purposes of war, or 'the art of military command', military commanders are the main actors. However, if understanding strategy as the use of the war for the purposes of policy, or 'the art of controlling and utilizing the resources of a nation', politicians become the object of study.[10]

A third important question concerns the relationship between strategy in theory and strategy in practice. Strategic Studies has developed in close relationship to practice. Indeed, one of the founders of Strategic Studies as an academic discipline, Bernard Brodie, called strategic theory 'a theory for action'.[11] However, with the development of strategy as a field of study

---

[5]Hew Strachan, 'Strategy in Theory; Strategy in Practice', this issue.
[6]Richard K. Betts, 'Is Strategy an Illusion?', *International Security*, 25/2 (2000), 5–50, 6.
[7]Jeffrey W. Meiser, 'Ends + Ways + Means = (Bad) Strategy', *Parameters* 46/4 (2016/2017), 81–91; Barry R. Posen, *The Sources of Military Doctrine: France, Britain, and Germany between the World Wars* (Ithaca: Cornell University Press 1984).
[8]Betts, 'Is Strategy an Illusion?', 6.
[9]Quoted in Baylis and Wirtz, 'Introduction', 4.
[10]Quotes from Edward Mead Earle, 'Introduction' in Edward Mead Earle (ed.), *Makers of Modern Strategy: Military Thought from Machiavelli to Hitler* (Princeton: Princeton University Press 1944), vii–xi, viii.
[11]Bernard Brodie, *War and Politics* (New York, NY: Macmillan Publishing 1973), 452.

outside of the military academies, the division between strategy in theory and strategy in practice has become larger. Already by the mid-1960s, Brodie admitted that strategic theory had drifted too far from the practice of strategy.[12] Several scholars have expressed similar concerns, and at the turn of the century, Betts noted that many academics do not 'grasp how hard it is to implement strategic plans'. Rationalist models of strategy, he argued, could only provide 'heuristic beginnings for real strategies which, by definition, must be demonstrably practical'.[13]

In the first article of this special issue, 'Strategy in Theory; Strategy in Practice', Hew Strachan contributes to this debate. According to Strachan, our understanding of strategy has changed over time. Indeed, he argues that the development of war since the end of the Cold War has left us especially uncertain of what strategy means, unclear about who makes strategy, and confused about the relationship between strategy in theory and strategy in practice.

In the days of Napoleon and Clausewitz, the focus of strategy was on how to win wars. But, when wars became more complex, strategy became increasingly connected to policy. In the seminal work, *Makers of Modern Strategy: Military Thought from Machiavelli to Hitler*, Edward Mead Earle argues that 'as war and society have become more complicated [...] strategy has of necessity required increasing consideration of nonmilitary factors, economic, psychological, moral, political, and technological'.[14] With the introduction of nuclear weapons in the 1940s, and the increasing focus on deterrence during the Cold War, strategy became about preventing war rather than waging it. In the event of nuclear war, experiences of traditional wars were not considered important, and the use of the battle for the purposes of war became all the more distant.

Since the end of the Cold War, and especially after 9/11, Strachan argues that 'the actual experience of war has required us to re-integrate [war and strategy] in ways that had not been necessary when war was more a threat than an actuality'.[15] This has made us confused. In the absence of strategy in the wars in Afghanistan and Iraq, the military has presented counterinsurgency doctrine as strategy, rather than the tactical method it is. At the same time, politicians have become all the more involved in tactical solutions to strategic problems, for example, through the use of drones for targeting enemy leaders. As a solution, Strachan suggests that the debate needs to 'be informed by the recognition of the distinction between strategy in theory and strategy in practice'.[16] He argues that both perspectives are

---

[12]Brodie, *War and Politics*, 474–475. For an overview of the development of Strategic Studies, see Richard K. Betts, 'Should Strategic Studies Survive?', *World Politics* 50 (1997), 7–33.
[13]Betts, 'Is Strategy an Illusion?', 7–8. See also Alexander L. George, *Bridging the Gap: Theory and Practice in Foreign Policy* (Washington, DC: United States Institute of Peace Press 1993); Joseph S. Nye, 'Bridging the Gap between Theory and Policy', *Political Psychology* 29/4 (2008), 593–603.
[14]Earle, 'Introduction', viii.
[15]Strachan, 'Strategy in Theory'.
[16]*Ibid.*

required, but needs to be related to each other. He concludes by stressing the more pragmatic aspects of strategy. Strategy, he argues, 'needs to be modest about itself and about what it can deliver. It is, after all, more of an art than a science, and it behoves those who think about it and those who practice it not to be too brazen about its status'.[17]

Strategic theory

Apart from questions about the discipline itself, one of the most central questions in Strategic Studies is how to use force or the threat of force to achieve desired ends. Different strategies of coercion, such as deterrence, compellence and coercive diplomacy, have been especially scrutinised.[18] While traditional strategic theory has primarily focused on the military strategy of states,[19] changes in international security and the conduct of war over the last decades have opened up new areas for the study of strategy.

One such area is peacekeeping operations. Peacekeeping is rarely considered a military endeavour. But, while traditional peacekeeping operations were only deploying a few number of troops with a limited mandate to use force, contemporary operations are usually large, with up to 20,000 troops, and with much more forceful mandates than before. These developments have made peacekeeping an increasingly important area for the study of military strategy. So far, this has largely been overlooked.[20]

---

[17]*Ibid.*

[18]See, for example, Daniel Byman and Matthew Waxman, *The Dynamics of Coercion: American Foreign Policy and the Limits of Military Might* (Cambridge: Cambridge University Press 2002); Lawrence Freedman, *Deterrence* (Cambridge: Polity Press 2004); Alexander L. George, *Forceful Persuasion: Coercive Diplomacy as an Alternative to War* (Washington, DC: United States Institute of Peace Press 1991); Peter Viggo Jakobsen, *Western Use of Coercive Diplomacy after the Cold War: A Challenge for Theory and Practice* (Houndmills, Basingstoke, Hampshire: Macmillan Press 1998); Glenn H. Snyder, *Deterrence and Defense: Toward a Theory of National Security* (Princeton, NJ: Princeton University Press 1961).

[19]Strategic Studies has indeed often been accused of being state centric, not including non-state actors, transnational groups and international organisations. See, for example, Baylis and Wirtz, 'Introduction', 11; Barry Buzan and Lene Hansen, *The Evolution of International Security Studies* (Cambridge: Cambridge University Press 2012), 37; Isabelle Duyvesteyn and James E. Worrall, 'Global Strategic Studies: A Manifesto', *Journal of Strategic Studies* 40/3 (2017), 347–357, 349; Keith Krause and Michael C. Williams, 'From Strategy to Security: Foundations of Critical Security Studies', in Keith Krause and Michael C. Williams (eds.), *Critical Security Studies: Concepts and Cases* (London: Routledge 1997), 33–59. See Pascal Vennesson for a refutation. Pascal Vennesson, 'Is Strategic Studies Narrow? Critical Security and the Misunderstood Scope of Strategy', *Journal of Strategic Studies* 40/3 (2017), 358–391, 368–372.

[20]Notable exceptions are, Alexander J. Bellamy, 'Lessons Unlearned: Why Coercive Diplomacy Failed at Rambouillet', *International Peacekeeping* 7/2 (2000), 95–114; Ken Ohnishi, 'Coercive Diplomacy and Peace Operations: Intervention in East Timor', *NIDS Journal of Defense and Security* 13 (2012), 53–77. For a special focus on the protection of civilians, see Arthur J. Boutellis, 'From Crisis to Reform: Peacekeeping Strategies for the Protection of Civilians in the Democratic Republic of the Congo', *Stability: International Journal of Security and Development* 2/3 (2013), 1–11; Stian Kjeksrud, 'The Utility of Force for Protecting Civilians', in Haidi Willmot, Ralph Mamiya, Scott Scheeran and Marc Weller (eds.), *Protection of Civilians* (Oxford: Oxford University Press 2016), 329–349; Paul D. Williams, *Enhancing Civilian Protection in Peace Operations: Insights from Africa* (Washington, DC: Africa Center for Strategic Studies 2010).

In the second article of this issue, 'Military Strategy and Peacekeeping: An Unholy Alliance?', Kersti Larsdotter addresses this question in a comprehensive manner. She outlines the logic of four main strategies—defence, deterrence, compellence and offence—for peacekeeping operations, using traditional strategic theory as a point of departure. She argues that all four can indeed be utilised for the most common strategic objectives in peacekeeping. Defensive strategies can be used in the protection of civilians as well as for self-defence. Deterrence can also be used for these purposes, but it can also be used to deter violence against the political process in general. Compellence is useful when the peacekeepers are deployed in an ongoing conflict, to stop violence against civilians or the disarmament process. Lastly, offensive strategies, although furthest from the peacekeeping norm, can deprive the spoilers of the means of continued fighting.

Larsdotter also traces the use of these strategies in two consecutive UN operations in the Democratic Republic of the Congo: MONUC and MONUSCO. She finds that all four strategies are indeed used in the two operations. They are, however, neither comprehensive nor proactive, leaving the true potential of military strategy unrealised. She concludes that while the military strategy of peacekeeping operations is ultimately restrained by a great number of factors, such as the need for consensus in the Security Council or the will of the troop contributing countries, it 'does not reduce the importance of having a logically coherent idea about how military force can contribute to keep the peace'.[21]

Another, quickly emerging, area for the study of strategy is cyber warfare.[22] The increasing dependence on information systems has made cyberspace a new 'war fighting domain'.[23] Issues such as the risk of cyber war,[24] and the consequences of the offensive and defensive abilities of cyber tools for the international system, i.e., the offense-defence balance, are commonly discussed.[25] Lately, scholars have increasingly turned their attention to cyber strategies, especially different forms of coercion, adapting

---

[21]Kersti Larsdotter, 'Military Strategy and Peacekeeping: An Unholy Alliance?', this issue.
[22]See, for example, the roundtable in Timothy J. Junio, 'How Probable Is Cyber War? Bringing IR Theory Back in to the Cyber Conflict Debate', *Journal of Strategic Studies*, 36/1 (2013), 125–33.
[23]Jon R. Lindsay, and Erik Gartzke, 'Coercion through Cyberspace: The Stability-Instability Paradox Revisited', in Kelly M. Greenhill and Peter Krause (eds.), *Coercion: The Power to Hurt in International Politics* (Oxford: Oxford University Press 2018), 179–203.
[24]Adam Liff, 'Cyberwar: A New "Absolute Weapon"? The Proliferation of Cyberwarfare Capabilities and Interstate War', *Journal of Strategic Studies* 35/3 (2012), 401–428; Thomas Rid, 'Cyber War Will Not Take Place', *Journal of Strategic Studies* 35/1 (2012), 5–32; Brandon Valeriano and Ryan C Maness, 'The Dynamics of Cyber Conflict between Rival Antagonists', *Journal of Peace Research* 51/3 (2014), 347–360. See also Timothy J. Junio, 'How Probable is Cyber War? Bringing IR Theory Back In to the Cyber Conflict Debate', *Journal of Strategic Studies* 36/1 (2013), 125–133, as well as the rest of the roundtable in the same issue of *Journal of Strategic Studies*.
[25]Ilai Saltzman, 'Cyber Posturing and the Offense-Defense Balance', *Contemporary Security Policy* 34/1 (2013), 40–63; Rebecca Slayton, 'What is the Cyber Offense-Defense Balance? Conceptions, Causes, and Assessment', *International Security* 41/3 (2016/2017), 72–109.

traditional strategic theory to this domain.[26] Although some scholars argue that cyber operations have limited coercive value, others argue that there are ways in which cyber operations can contribute to successful coercion, for example, by imposing costs and destabilising an opponent's leadership,[27] or by supplementing conventional deterrence strategies.[28]

In the third article of this special issue, 'Fancy Bears and Digital Trolls: Cyber Strategy with a Russian Twist', Benjamin Jensen, Brandon Valeriano and Ryan Maness outline the logic of cyber coercion. They argue that cyber operations are similar to covert action. Therefore, states cannot clearly communicate their intents through cyber threats, making it a rather weak coercive tool. It should therefore not be used in isolation, but as an additive measure to amplify existing strategic signals. It should also be used to shape long-term competition, rather than seek direct concessions. Based on these assumptions, Jensen et al. propose three strategies for cyberspace: disruption, espionage and degradation. Disruption is used to shape the larger bargaining context, espionage to alter the balance of information to achieve a position of advantage, and degradation to sabotage the enemy target's network, operation or systems.

Whereas much research on cyber coercion lacks empirical analysis,[29] Jensen et al. use their framework to analyse two cases of documented cyber operations pursued by Russia: the manipulation of the 2016 US election, and cyber operations in Ukraine during ongoing conflict. They find that while Russia's operations against the US election follow classic tactics for Russian disinformation campaigns, their operations in Ukraine is a new style of information operations designed to obfuscate. Russian cyber strategy, they argue, seems to be more about ambiguous signalling and amplifying propaganda, rather than achieving direct political concessions. They conclude that if 'Russian cyber operations are a harbinger of the future of strategy, they show a world prone to intrigue and political warfare more than they do violent military escalation and interstate conflict'.[30]

---

[26]See, for example, Brian M. Mazanec and Bradley A. Thayer, *Deterring Cyber Warfare: Bolstering Strategic Stability in Cyberspace* (Houndmills, Basingstoke, Hampshire: Palgrave Macmillan 2015); Travis Sharp, 'Theorizing Cyber Coercion: The 2014 North Korean Operation against Sony', *Journal of Strategic Studies* 40/7 (2017), 898–926; Uri Tor, '"Cumulative Deterrence" as a New Paradigm for Cyber Deterrence', *Journal of Strategic Studies* 40/1–2 (2017), 92–117.

[27]Sharp, 'Theorizing Cyber Coercion'.

[28]Michael Fischerkeller, 'Incorporating Offensive Cyber Operations into Conventional Deterrence Strategies', *Survival* 59/1 (2017), 103–134.

[29]Some exceptions are, Rid, Thomas, *Cyber War Will Not Take Place* (Oxford: Oxford University Press 2013); Sharp, 'Theorizing Cyber Coercion'; Valeriano and Maness, 'The Dynamics of Cyber Conflict between Rival Antagonists'; Brandon Valeriano and Ryan C. Maness, *Cyber War versus Cyber Realities: Cyber Conflict in the International System* (Oxford: Oxford University Press 2015).

[30]Ben Jensen, Brandon Valeriano and Ryan Maness, 'Fancy Bears and Digital Trolls: Cyber Strategy with a Russian Twist', this issue.

The makers of strategy

In the last section of this special issue, focus is turned to the makers of strategy. This area of study includes several sets of research questions, for example, who are the strategy makers and how do they make strategy? One of the more debated issues is why states pursue the strategies they do. While many states choose strategies that seems rational for pursuing their political goals, other states choose strategies that seems suboptimal or very poor. Why do, for example, states pursue strategies of military intervention despite repeated failures to translate tactical success on the battlefield into meaningful political resolutions? Or how can we explain that small states, with limited military capability, pursue strategies of escalation against much larger and more powerful states?

Some scholars have explored factors internal to military organisations, such as military culture or bureaucratic forces.[31] One common internal explanation for suboptimal strategies is the civil-military dynamic in the making of strategy. Risa Brooks, for example, argues that preference divergence and the balance of power between military and political leaders affect the quality of strategic assessments, which in turn influence both the formulation and execution of strategies.[32]

In the fourth article of this special issue, 'The Political-Military Dynamic in the Conduct of Strategy', John Kiszely argues that 'the constructive and effective interaction between politicians [...] and their military advisers' is key to successful grand strategy, but that this civil-military dynamic is especially problematic in the twenty-first century.[33] He identifies several reasons for why this is the case. Some of the problems are due to changes in the character of war and conflict. The decreased distance between the strategic and tactical levels in war makes the differences in perception between politicians and the armed forces more acute, and confuses the right level of decision. What seems like tactical level decisions for the military might be considered major strategic decisions for the politicians, increasing the friction between the two.[34] The inclusion of non-military lines of operations (diplomatic, economic and social), and the long and unpredictable nature of non-state wars, also makes the civil-military dynamic more

---

[31]See, for example, Elizabeth Kier, *Imagining War: French and British Military Doctrine between the Wars* (Princeton, NJ: Princeton University Press 1997); John A. Nagl, *Eating Soup with a Knife: Counterinsurgency Lessons from Malaya and Vietnam* (Chicago: University of Chicago Press 2005); Posen, *The Sources of Military Doctrine*; Jack Snyder, *The Ideology of the Offensive: Military Decision Making and the Disasters of 1914* (Ithaca: Cornell University Press 1984).

[32]Risa Brooks, *Shaping Strategy: The Civil-Military Politics of Strategic Assessment* (Princeton: Princeton University Press 2008).

[33]John Kiszely, 'The Political-Military Dynamic in the Conduct of Strategy', this issue.

[34]On the 'strategic corporal', see, for example, Charles C. Krulak, 'The Strategic Corporal: Leadership in the Three Block War', *Marines Magazine* (January 1999); Chiara Ruffa, Christopher Dandeker and Pascal Vennesson, 'Soldiers Drawn into Politics? The Influence of Tactics in Civil-Military Relations', *Small Wars & Insurgencies* 24/2 (2013), 322–334.

problematic. In addition, Kiszely suggests that there is an increasing erosion of trust between political decision-makers and their senior military advisers, enhancing the problems of civil-military relations further. On the one hand, politicians have less direct military experience than a century ago, and they are usually lacking training in strategic theory. Contemporary senior military officers, on the other hand, tend to use media to leverage public support for their cause, in order to increase their influence on political decisions.

To make the political-military dynamic more successful, Kiszely proposes several remedies. He suggests, for example, that both politicians and military officers need a better understanding of strategy and the interplay between war and politics. This is supported by Strachan in his contribution to this special issue. Since practitioners at that level are extremely occupied, Kiszely argues that this education needs to take place early in their careers.[35] He also suggests that politicians and military officers need to find greater acceptance and understandings for each other's domains. Military officers need to accept that politics is part of war, and politicians need to acknowledge their need of military expertise in the balancing of ends, ways and means. He concludes that a pre-requisite for states to make the right choices and develop optimal strategies is to recognise that the political-military dynamic is inherently problematic and that mutual understanding, respect and trust are at its foundation.

Another set of explanatory factors for why states pursue suboptimal strategies, are external of the military organisations, for example, the conflict environment or domestic audiences. According to, for example, Barry R. Posen, shifts in the balance of power can explain the strategic choices of states. In his study on the sources of military doctrine, where he compares the explanatory power of organisational theory and the balance of power theory, he concludes that 'when threats become sufficiently grave, soldiers themselves begin to reconsider organizationally self-serving doctrinal preferences, if those preferences do not adequately respond to the state's immediate security problems'.[36] The cost-benefit calculation of domestic audiences is also a common explanation for suboptimal strategies, at least in the case of democratic states. Since wars are usually highly costly endeavours, leaders chose strategies that minimises losses, even when these strategies are not optimal for winning wars.[37]

Whereas previous research primarily focuses on either internal or external constraints on one level only, Michael Mayer makes use of a comprehensive

---

[35]This is supported by Nye, 'Bridging the Gap between Theory and Policy', 600.

[36]Posen, *The Sources of Military Doctrine*, 239–240. Fravel argues that for less developed states, it is rather changes in the conduct of *warfare* in the international system that influence the strategic choices of states. M. Taylor Fravel, 'Shifts in Warfare and Party Unity: Explaining China's Changes in Military Strategy', *International Security* 42/3 (2017/2018), 37–83.

[37]Jonathan D. Cavalery, 'The Myth of Military Myopia: Democracy, Small Wars, and Vietnam', *International Security* 34/3 (2009/2010), 119–157.

approach in his contribution to this special issue, 'Trigger Happy: The Foundations of US Military Interventions'. He suggests that both internal and external factors at several levels together can explain the United States' suboptimal strategy of interventions, i.e., its strategy of persistent military interventions in other states despite meagre prospects for successful outcomes and questionable strategic benefits.

Changes at the systemic level have made the strategy of intervention more conducive. The United States' position as the sole super power after the end of the Cold War has eliminated the risk of triggering a great power conflict when intervening in conflicts elsewhere. At the same time, the increasing levels of sub-state threats in the international system have increased the motivation for intervention. At the state level, the growing power and influence of the executive branch, at the expense of Congress, have allowed for more insulated decision-making processes, focused on short-term fixes rather than long-term solutions. Additionally, the militarisation of US foreign policy has made military solutions to perceived threats the norm. Finally, domestic societal factors also encourage interventions. Although culturally attuned to the use of force, Mayer argues, the population of the United States is highly sensitive to fatalities, making limited military interventions the obvious choice. Furthermore, a disengaged and polarised public fail to evaluate the utility of interventions properly. Mayer concludes that even though 'the United States has a long history of military interventionism, political and institutional developments have made it easier to choose military force as a foreign policy tool, leading to a persistent pattern of misapplication of armed force'.[38]

In the last article of this issue, 'Weak Party Escalation: An Underestimated Strategy for Small States?', Jan Angstrom and Magnus Petersson aim to explain another case of suboptimal strategies, i.e., why weak states escalate against much larger powers. Although not particularly common, weak-party escalation does occur, and the actions of the North Korean leader Kim Jong-un against the United States in 2017 highlight our need to know more about why weak parties escalate against stronger parties. While escalation is certainly not a new issue in Strategic Studies, most theories on escalation in interstate wars are only including relatively equal opponents.[39] Weak-party escalation has rather been understood as 'mistakes emanating from misperceptions or downright madness'.[40] Instead of using internal or external

---

[38]Michael Mayer, 'Trigger Happy: The Foundations of US Military Interventions', this issue.

[39]Bernard Brodie, *Escalation and the Nuclear Option* (Princeton, NJ: Princeton University Press 1966); Herman Kahn, *On Escalation: Metaphors and Scenarios* (Baltimore, MD: Penguin 1965); Barry R. Posen, *Inadvert Escalation: Conventional War and Nuclear Risks* (Ithaca, NY: Cornell University Press 1991); Thomas C. Schelling, *Arms and Influence* (New Haven: Yale University Press 2008); Richard Smoke, *War: Controlling Escalation* (Cambridge, MA: Harvard University Press 1977).

[40]Jan Angstrom and Magnus Petersson, 'Weak Party Escalation: An Underestimated Strategy for Small States', this issue.

factor to explain why states pursue poor strategies, however, Angstrom and Petersson argue that weak-party escalation is actually not always irrational.

Angstrom and Petersson outline the logic of four strategies of weak-party escalation in state-on-state relations, and they suggest that there are two situations when it is rational for weak parties to escalate against stronger adversaries. First, it becomes rational when the goal of escalation is not to threaten the stronger adversary directly. Theories of escalation rest on the logic of coercion rather than brute force, and involve the threat of future damage. One of the prevailing ideas is that of escalation dominance, i.e., that the party that have the 'ability to inflict more costs on an adversary than the adversary can inflict in return', will win.[41] Hence, if weak parties escalate to achieve a position to hurt their opponent directly, escalation becomes an irrational strategy. But, if the weak party want to *provoke* a desired over-reaction from the stronger adversary to trigger outside help, or to forge a *reputation* of not yielding lightly, as a way of future, long-term, deterrence, weak-party escalation becomes rational. Second, weak-party escalation is also rational when the weak party can create some sort of escalation dominance, either by *compartmentalizing* the conflict within one domain in which it has escalation dominance or by creating *a division of labour* with a stronger ally.

Following this escalation logic, it becomes evident that weak parties, especially small states, do not need to be '"passive victims" in international affairs, totally in the hands of great power competition and power play'.[42] Instead, policy-makers in weaker-party relations have several strategies to choose between to achieve their strategic goals. While Angstrom and Petersson warn for several dangers associated with weak-party escalation, for example, that possible outside help is jeopardised if the weak party is being regarded as the aggressor when escalating to provoke an over-reaction from a stronger adversary, they conclude 'that leaders of small states should realise that resistance and escalation, even against great powers, can be beneficial'.[43]

The future of military strategy

In 1949, Bernard Brodie called for the development of strategy as a systematic field of study. He argued that 'strategy is not receiving the scientific treatment it deserves either in the armed services or, certainly, outside of them'.[44] Seventy years later, students of strategy can be found

---

[41]Byman and Waxman, *The Dynamics of Coercion*, 38–39.
[42]Angstrom and Petersson, 'Weak Party Escalation'.
[43]*Ibid.*
[44]Bernard Brodie, 'Strategy as a Science', *World Politics* 1/4 (1949), 467–488, 468.

in universities and military academies around the world, and various aspects of strategy are fervently debated.

The articles in this special issue are contributing to this debate in mainly three ways. First, they enhance our knowledge of how states and international organisations can use force in contemporary conflicts. Strategies of offense, defence and coercion, for example, can all be used to achieve various strategic aims in peacekeeping operations. These strategies need, however, to be comprehensive and proactive in nature. Coercion can also be useful in the cyber domain. Because of the covert nature of cyber operations, these operations should be used as an additive measure to amplify existing strategic signals, with the aim of shaping long-term competition rather than seek direct concessions. Small states, usually seen as passive victims in the international system, also have more strategic choices than previously appreciated. When the conditions are right, it is rational for small states to use escalation against more powerful states.

Second, these articles also contribute to our knowledge of how to avoid suboptimal choices in the making of strategy. Because of the problems in the civil-military dynamic, enhanced by the changes of war in the twenty-first century, building trust and acceptance between civilian and military makers of strategy is especially important. The changes of war and international security also affect the environment in which strategy making takes place, and we need to be observant of how factors at various levels, including the systemic, state, societal and domestic levels, influence our strategy making. However, it is also important to be mindful that certain strategies might be rational, even when they seem not to be.

Finally, several authors in this special issue have highlighted the relation between strategy in theory and strategy in practice. The study of strategy is important for the practice of strategy. Not only can it contribute to a better understanding of how to use force in the international system, potentially contributing to better strategies. It is also valuable in the education of civilian and military practitioners. By creating a better understanding and acceptance of each other's domains, Strategic Studies can improve the civil-military dynamics in the making of strategy. But, it is also important to remember that while strategy in theory needs to inform strategy in practice, it is crucial to make a distinction between the two.

Disclosure statement

No potential conflict of interest was reported by the author.

ORCID

Kersti Larsdotter ⓘ http://orcid.org/0000-0001-5843-3878

Bibliography

Angstrom, Jan and Magnus Petersson, 'Theorizing Weak-Party Escalation in Asymmetric Conflicts', (this issue).

Baylis, John and J. Wirtz James, 'Introduction', in John Baylis, James Wirtz, Eliot Cohen, and Colin Gray (eds.), Strategy in the Contemporary World: An Introduction to Strategic Studies (Oxford: Oxford University Press 2002), 1–14.

Baylis, John, James Wirtz, Eliot Cohen, and Colin Gray, (eds.), *Strategy in the Contemporary World: An Introduction to Strategic Studies* (Oxford: Oxford University Press 2002).

Bellamy, Alexander J., 'Lessons Unlearned: Why Coercive Diplomacy Failed at Rambouillet', *International Peacekeeping*, 7/2 (2000), 95–114.

Betts, Richard K., 'Should Strategic Studies Survive?' *World Politics*, 50 (1997), 7–33. doi:10.1017/S0043887100014702

Betts, Richard K., 'Is Strategy an Illusion?' *International Security*, 25/2 (2002), 5–50.

Bird, Tim and Alex Marshall, *Afghanistan: How the West Lost Its Way* (New Haven, CT: Yale University Press 2011).

Boutellis, Arthur J., 'From Crisis to Reform: Peacekeeping Strategies for the Protection of Civilians in the Democratic Republic of the Congo', *Stability: International Journal of Security and Development*, 2/3 (2013), 1–11.

Brodie, Bernard, 'Strategy as a Science', *World Politics*, 1/4 (1949), 467–88. doi:10.2307/2008833

Brodie, Bernard, *Escalation and the Nuclear Option* (Princeton, NJ: Princeton University Press 1966).

Brodie, Bernard, *War and Politics* (New York, NY: Macmillan Publishing 1973).

Brooks, Risa, *Shaping Strategy: The Civil-Military Politics of Strategic Assessment* (Princeton: Princeton University Press 2008).

Buzan, Barry and Lene Hansen, *The Evolution of International Security Studies* (Cambridge: Cambridge University Press 2012).

Byman, Daniel and Matthew Waxman, *The Dynamics of Coercion: American Foreign Policy and the Limits of Military Might* (Cambridge: Cambridge University Press 2002).

Cavalery, Jonathan D., 'The Myth of Military Myopia: Democracy, Small Wars, and Vietnam', *International Security*, 34/3 (2009/2010), 119–57.

Clausewitz, Carl von, *On War* translated and edited by Michel Howard and Peter Paret (Princeton, NJ: Princeton University Press 1976).

Duyvesteyn, Isabelle and James E. Worrall, 'Global Strategic Studies: A Manifesto', *Journal of Strategic Studies*, 40/3 (2017), 347–57.

Earle, Edward Mead, 'Introduction', in Edward Mead Earle (ed.), Makers of Modern Strategy: Military Thought from Machiavelli to Hitler (Princeton: Princeton University Press 1944), vii–xi.

Finlan, Alastair, *Contemporary Military Strategy and the Global War on Terror: US and UK Armed Forces in Afghanistan and Iraq 2001-2012* (London: Bloomsbury Academic 2014).

Fischerkeller, Michael, 'Incorporating Offensive Cyber Operations into Conventional Deterrence Strategies', *Survival*, 59/1 (2017), 103–34. doi:10.1080/00396338.2017.1282679

Fravel, M. Taylor, 'Shifts in Warfare and Party Unity: Explaining China's Changes in Military Strategy', *International Security*, 42/3 (2017/2018), 37–83.

Freedman, Lawrence, *Deterrence* (Cambridge: Polity Press 2004).

George, Alexander L., *Forceful Persuasion: Coercive Diplomacy as an Alternative to War* (Washington, DC: United States Institute of Peace Press 1991).

George, Alexander L., *Bridging the Gap: Theory and Practice in Foreign Policy* (Washington, DC: United States Institute of Peace Press 1993).

Jakobsen, Peter Viggo, *Western Use of Coercive Diplomacy after the Cold War: A Challenge for Theory and Practice* (Houndmills, Basingstoke, Hampshire: Macmillan Press 1998).

Jensen, Ben, Brandon Valeriano, and Ryan Maness, 'Fancy Bears and Digital Trolls: Cyber Strategy with a Russian Twist', (this issue).

Junio, Timothy J., 'How Probable Is Cyber War? Bringing IR Theory Back in to the Cyber Conflict Debate', *Journal of Strategic Studies*, 36/1 (2013), 125–33. doi:10.1080/01402390.2012.739561

Kahn, Herman, *On Escalation: Metaphors and Scenarios* (Baltimore, MD: Penguin 1965).

Kane, Thomas M. and David J. Lonsdale, *Understanding Contemporary Strategy* (London: Routledge 2012).

Kier, Elizabeth, *Imagining War: French and British Military Doctrine between the Wars* (Princeton, NJ: Princeton University Press 1997).

Kiszely, John, 'The Political-Military Dynamic in the Conduct of Strategy', *Journal of Strategic Studies*, (this issue), 1–24. doi:10.1080/01402390.2018.1497488

Kjeksrud, Stian, 'The Utility of Force for Protecting Civilians', in Haidi Willmot, Ralph Mamiya, Scott Scheeran, and Marc Weller (eds.), *Protection of Civilians* (Oxford: Oxford University Press 2016), 329–49.

Krause, Keith and Michael C. Williams, 'From Strategy to Security: Foundations of Critical Security Studies', in Keith Krause and Michael C. Williams (eds.), Critical Security Studies: Concepts and Cases (London: Routledge 1997), 33–59.

Krulak, Charles C., 'The Strategic Corporal: Leadership in the Three Block War', *Marines Magazine* (January 1999).

Larsdotter, Kersti, 'Military Strategy and Peacekeeping: An Unholy Alliance?' (this issue).

Liff, Adam, 'Cyberwar: A New "Absolute Weapon"? the Proliferation of Cyberwarfare Capabilities and Interstate War', *Journal of Strategic Studies*, 35/3 (2012), 401–28. doi:10.1080/01402390.2012.663252

Lindsay, Jon R. and Erik Gartzke, 'Coercion through Cyberspace: The Stability-Instability Paradox Revisited', in Kelly M. Greenhill and Peter Krause (eds.),

Coercion: The Power to Hurt in International Politics (Oxford: Oxford University Press 2018), 179–203.

Mahnken, Thomas G. and Joseph A. Maiolo, (eds.), *Strategic Studies: A Reader* (London: Routledge 2008).

Mayer, Michael, 'Trigger Happy: The Foundations of US Military Interventions', (this issue).

Mazanec, Brian M. and Bradley A. Thayer, *Deterring Cyber Warfare: Bolstering Strategic Stability in Cyberspace* (Houndmills, Basingstoke, Hampshire: Palgrave Macmillan 2015).

Meiser, Jeffrey W., 'Ends + Ways + Means = (Bad) Strategy', *Parameters*, 46/4 (2016/ 2017), 81–91.

Nagl, John A., *Eating Soup with a Knife: Counterinsurgency Lessons from Malaya and Vietnam* (Chicago: University of Chicago Press 2005).

Nye, Joseph S., 'Bridging the Gap between Theory and Policy', *Political Psychology*, 29/4 (2008), 593–603.

Ohnishi, Ken, 'Coercive Diplomacy and Peace Operations: Intervention in East Timor', *NIDS Journal of Defense and Security*, 13 (2012), 53–77.

Posen, Barry R., *The Sources of Military Doctrine: France, Britain, and Germany between the World Wars* (Ithaca: Cornell University Press 1984).

Posen, Barry R., *Inadvert Escalation: Conventional War and Nuclear Risks* (Ithaca, NY: Cornell University Press 1991).

Rid, Thomas, 'Cyber War Will Not Take Place', *Journal of Strategic Studies*, 35/1 (2012), 5–32. doi:10.1080/01402390.2011.608939

Rid, Thomas, *Cyber War Will Not Take Place* (Oxford: Oxford University Press 2013).

Ruffa, Chiara, 'Christopher Dandeker and Pascal Vennesson, 'Soldiers Drawn into Politics? The Influence of Tactics in Civil-Military Relations'', *Small Wars & Insurgencies*, 24/2 (2013), 322–34. doi:10.1080/09592318.2013.778035

Saltzman, Ilai, 'Cyber Posturing and the Offense-Defense Balance', *Contemporary Security Policy*, 34/1 (2013), 40–63. doi:10.1080/13523260.2013.771031

Schelling, Thomas C., *Arms and Influence* (New Haven: Yale University Press 2008).

Sharp, Travis, 'Theorizing Cyber Coercion: The 2014 North Korean Operation against Sony', *Journal of Strategic Studies*, 40/7 (2017), 898–926. doi:10.1080/ 01402390.2017.1307741

Slayton, Rebecca, 'What Is the Cyber Offense-Defense Balance? Conceptions, Causes, and Assessment', *International Security*, 41/3 (2016/2017), 72–109.

Sloan, Elinor C., *Modern Military Strategy: An Introduction* (London: Routledge 2017).

Smoke, Richard, *War: Controlling Escalation* (Cambridge, MA: Harvard University Press 1977).

Snyder, Glenn H., *Deterrence and Defense: Toward a Theory of National Security* (Princeton, NJ: Princeton University Press 1961).

Snyder, Jack, *The Ideology of the Offensive: Military Decision Making and the Disasters of 1914* (Ithaca: Cornell University Press 1984).

Strachan, Hew, 'Strategy or Alibi? Obama, McChrystal and the Operational Level of War', *Survival*, 52/5 (2010), 157–82. doi:10.1080/00396338.2010.522104

Strachan, Hew, 'Strategy in Theory; Strategy in Practice', (this issue).

Tor, Uri, '"Cumulative Deterrence" as a New Paradigm for Cyber Deterrence', *Journal of Strategic Studies*, 40/1–2 (2017), 92–117. doi:10.1080/01402390.2015.1115975

Valeriano, Brandon and Ryan C. Maness, 'The Dynamics of Cyber Conflict between Rival Antagonists', *Journal of Peace Research*, 51/3 (2014), 347–60. doi:10.1177/ 0022343313518940

Valeriano, Brandon and Ryan C. Maness, *Cyber War versus Cyber Realities: Cyber Conflict in the International System* (Oxford: Oxford University Press 2015).

Vennesson, Pascal, 'Is Strategic Studies Narrow? Critical Security and the Misunderstood Scope of Strategy', *Journal of Strategic Studies*, 40/3 (2017), 358–91. doi:10.1080/01402390.2017.1288108

Williams, Paul D., *Enhancing Civilian Protection in Peace Operations: Insights from Africa* (Washington, DC: Africa Center for Strategic Studies 2010).

Strategy in theory; strategy in practice

Hew Strachan

ABSTRACT
The practice of strategy is different from strategic theory. The latter was largely developed by professional soldiers from the experiences of the Napoleonic Wars, and compared the present with the past to establish general truths about war. It used history as its dominant discipline until 1945. The advent of nuclear weapons made history seem less relevant, and prompted the inclusion of other disciplines; deterrence theory also made strategic theory more abstract and distant from the practice of war. Since 9/11, the experience of war has forced strategy to become less theoretical and to do better in reconciling theory with practice.

In his book, *Modern Strategy*, published in 1999, Colin Gray declared with his customary forcefulness (and the italics are his): '*there is an essential unity to all strategic experience in all periods of history because nothing vital to the nature and function of war and strategy changes*'.[1] Gray is not alone among scholars working on the place of war in international relations who see history as a continuum. Edward Luttwak, Beatrice Heuser and Lawrence Freedman have taken similar lines, albeit with less directness. Nor are they wrong to do so: it is better that history contributes to an understanding of war and strategy than it does not. However, history is not just a repository from which we cherry-pick enduring truths. If it were, there would be little value in obeying Michael Howard's injunction that we study military history in width, depth and context.[2] The more we do that, the more we see nuance, difference, and even discontinuity.

Colin Gray is a student of politics with a strong interest in history; this essay is written by a historian with a strong interest in policy. The value that Gray sees in history is a sense of continuity. Many historians, particularly

---

[1]Colin S. Gray, *Modern Strategy* (Oxford: Oxford University Press 1999), 1. Colin Gray has responded to the criticism which follows in *The Strategy Bridge: Theory for Practice* (Oxford: Oxford University Press 2010), 10. Similar assumptions underpin John Lewis Gaddis, *On grand strategy* (London: Allen Lane 2018).
[2]Michael Howard, *The Causes of Wars and Other Essays* (London: Temple Smith 1983), 195–197.

those who look to the *longue durée*, set out with a similar purpose. There is no more telling or acerbic critic of those determined to describe a current development as unprecedented than the historian – who rightly says that there is nothing new about the rise of non-state actors in war, or about the incidence of civil war, or about the inter-weaving of crime and war. But what also attracts historians is the study of change: the causes of the French or Russian revolutions, the outbreak of the First World War, the end of the Cold War, the impact of the 9/11 attacks. Those who lived through those events were conscious of epoch-making change, and – however much we may soften their disruptive effects as we seek context and distance – we distort the past if we strive too zealously to minimise the impact of contingency or the effect of shock. Explaining change is both more difficult and more contentious (as testified by the enduring capacity of explanations for the major caesuras in world history to generate controversy) than accounting for continuity.

What we understand by strategy, in particular, has changed considerably over time. The word did not become current until the late eighteenth century and was not in regular use until after the final defeat of Napoleon in 1815. Napoleon himself did not employ it, at least during his active career: his critics might say it would have been to his advantage if he had.[3] However, that point itself illustrates the effect of change. Carl von Clausewitz, who was responding to the impact of Napoleon on war and who used his magnum opus *Vom Kriege* to develop the understanding which he derived from his own experience of Napoleonic warfare, regularly defined strategy as 'the use of the engagement for the purpose of the war'.[4] For him – as for Napoleon – battle lay at the centre of strategy, just as it did of war. But for modern tastes that approach is both too operational and insufficiently political. It defines strategy in ways that are narrowly military and therefore introspective. Decisive battle, as it was understood in the age of Frederick the Great or Napoleon, all but disappeared from warfare over the course of the twentieth century. The so-called battles of the First World War, of which Verdun and the Somme might stand as exemplars, lasted months and were not decisive. They had to be rationalised in new ways, through the vocabulary of attrition, a strategic idea which those wedded to the old forms of war vehemently rejected as nihilistic.[5]

That debate ought to have raised more profound and fundamental questions about the strategic theories which took battle's centrality for granted than it did. After the First World War, Basil Liddell Hart developed

[3]Napoléon, *De la guerre*, Presenté et annoté par Bruno Colson (Paris: Perrin 2011), 148.
[4]Carl von Clausewitz, *On War*, translated and edited by Michael Howard and Peter Paret (Princeton: Princeton University Press 1976), 128, 177.
[5]Friedrich von Bernhardi, *Vom Kriege der Zukunft. Nach der Erfahrungen des Weltkrieges* (Berlin: E.S. Mittler & Sohn 1920), 136–137.

a model of strategy which – in focusing on the 'indirect approach' – de-emphasised battle.[6] He went on, as he put it, to 're-frame' strategy by aligning it more closely with the use of military means to fulfil the ends of government.[7] In other words, strategy now looked outwards to its relationship with policy. Battle was not an end in itself, as it was for Napoleon, but a means to an end – and very often, and even increasingly, an optional one. This more modern understanding of strategy is that with which Colin Gray works. It is also fuels Napoleon's critics, who denounce him for failing to see that the purpose of war was not to create the conditions for the next battle but the capacity to convert war into a lasting peace. It behoves all those who read Clausewitz today in order to acquire a better comprehension of war to realise that, when he used the word 'strategy', he meant something very different from those who write about strategy today.

This distinction – strategy as the use of the battle for the purposes of war and strategy as the use of war for the purposes of policy – has become muddled, for perfectly understandable reasons. To be sure, titles like 'military' strategy on the one hand, and 'grand' or 'national' strategy on the other, convey the difference between the operational and political levels. British joint operational doctrine in 2004 stated that 'as the military component of strategy, military strategy is the process by which military objectives and force levels, which will assist in the achievement of political objectives, are decided'. It went on to stress that 'any document setting out a military strategy must contain an explanation of how the military strategy is to be integrated with other non-military elements of the national strategy'.[8] Although these distinctions reflect NATO doctrines, we are far from rigorous in their application, and understandably so. Steven Jermy, writing on strategy in 2011, included a heading for 'military strategy' in his index, but then said 'see politico-military strategy'.[9] He is not alone in his desire to reject a division between 'military' strategy and strategy more generally. John Stone, in a book published in the same year as Jermy's and explicitly titled *Military Strategy*, addressed not the process by which military objectives and force levels are set, but 'the instrumental link between military means and political ends'. For Stone the value of the epithet 'military' lay not in the distinction between the self-contained world of the soldier and the political level, which he saw as necessarily linked rather than bifurcated, but in the distinction between 'military strategy' and grand strategy, 'an activity that is

---

[6]Basil Liddell Hart, *The Decisive Wars of History: A Study in Strategy* (London: G. Bell & Sons Ltd 1929), which became *The Strategy of the Indirect Approach* (London: Faber & Faber 1942).

[7]Basil Liddell Hart, 'Strategy re-framed', in *When Britain Goes to War: Adaptability and Mobility* (London, Faber & Faber 1932).

[8]*Joint Doctrine Publication 01: Joint Operations* (Joint Doctrine and Concepts Centre, March 2004), para 208, 2-3–2-4.

[9]Steven Jermy, *Strategy for Action: Using Force Wisely in the 21st Century* (London: Knightstone Publishing 2011), 327.

concerned with the application of the totality of national resources in the pursuit of political goals'.[10] The multiplicity of meanings linked to 'military strategy' does not stop there. For example, some employ it to convey the distinction between strategy as an instrument of statecraft and its use in non-military contexts, particularly business.[11]

Paradoxically, the concepts which underpin 'grand strategy' may possess greater antiquity than do those around the operational meanings of strategy developed over the course of the 'long' nineteenth century. Pre-modern rulers may not have used the phrase 'grand strategy', but they certainly had to address the relationship between war and peace, and to understand the utility of military force within it. This is the justification for the inclusion of Thucydides's history of the Peloponnesian War on war college syllabuses, and for Edward Luttwak's determination to describe what he calls the 'grand strategies' of the Roman and Byzantine empires.[12] Historians of the Renaissance and early modern Europe have become increasingly comfortable with 'grand strategy' as a description of the policies pursued by their subjects: consider, for example, the work of Geoffrey Parker in looking at imperial Spain and the case of Philip II.[13] In that sense, it was the very invention of more modern 'operational' understandings of strategy by Clausewitz and his contemporary, Antoine-Henri Jomini, which first precipitated the confusion we now confront. The century-long effort to digest the impact of Napoleon coincided with the rise – not least intellectually – of the military profession and the notion of strategy as embedded within war was part and parcel of the self-validation of career officers.

This changed after 1945. The Charter of the United Nations had more to say about the role of international law in causing or forestalling war, *ius ad bellum*, than it did about the conduct of war, *ius in bello*. During the Cold War the threat of the use of war in order to preserve the global order – deterrence – became the principal function of many armed forces. Strategy in other words was as much, if not more, about preventing war as waging it. Strategy did indeed shape foreign policy, and foreign policy in turn was shaped disproportionately by strategic theory, as expressed in ideas like deterrence. Indeed, so theoretical was nuclear strategy, and so divorced from actual war, that from the late 1950s onwards it was largely developed by academics, most of them not historians.

---

[10]John Stone, *Military Strategy: The Politics and Technique of War* (London: Continuum, 2011), 4.

[11]Hervé Coutau-Bégarie, *Traité de stratégie* (6th edition, Paris: Economica 2008), 88–90.

[12]Edward N. Luttwak, *The Grand Strategy of the Roman Empire from the First Century A.D. to the Third* (1976) (London: Weidenfeld & Nicolson 1999); Edward N. Luttwak, *The Grand Strategy of the Byzantine Empire* (Cambridge MA: The Belknap Press of Harvard University Press 2009).

[13]Geoffrey Parker, *The Grand Strategy of Philip II* (New Haven, CT: Yale University Press 1998). This paragraph reflects a conference on 'Strategy and its making in early modern Europe', held at the University of St Andrews in honour of Geoffrey Parker, 29–30 April 2016.

In some respects, these changes can be seen as 'change back', a reversion to something that had existed before, as much as change *de novo*. Strategy re-acquired the power political connotations of which the military profession had robbed it in the years between Napoleon and Hitler. It ceased to be the monopoly of generals and military professionals, who became marginalised in its development. Since the Cold War's end, we have become confused about strategy not least because the actual experience of war has required us to re-integrate the two approaches in ways that had not been necessary when war was more a threat than an actuality. As a result, we are uncertain what strategy means and unclear who makes it. Is it the responsibility of its nineteenth-century protagonists, the armed forces, or of governments? Is war too important to be left to the generals, as France's prime minister in 1917–1918, Georges Clemenceau, opined? Or are they the only ones who – because of their life-long professional engagement – truly understand it, as Brigadier-General Jack D. Ripper insists in Stanley Kubrick's 1964 film, *Dr Strangelove, or how I learned to stop worrying and love the bomb*? Ripper says that Clemenceau may have been right 50 years ago, but 'now war is too important to be left to the politicians. They have neither the time, the training, nor the inclination for strategic thought'. Kubrick portrayed Ripper as insane: another 50 years on and George W. Bush and Tony Blair responded to the 9/11 attacks by invading Iraq, which had had no part in them, leading satirists to see statesmen, not soldiers, as war-mongers.

Or is strategy the task of a joint body like a national security council? The composition of such bodies varies from country to country according to national and constitutional norms, but it has the potential to combine 'military' strategy with wider approaches to strategy, because chiefs of staff either attend meetings or are members in their own right. It can also create 'grand' strategy because in some cases it integrates in a single body all ministers whose portfolios affect national security. The president or the prime minister is frequently in the chair. But if a national security council is the best institutional solution for the formation of strategy, and an increasing number of governments seem to think it is, why don't all states have one? Perhaps those which don't have either a national security strategy or a national security council are behaving with greater realism – precisely because they do not, in fact, intend to use war as an act of national policy, and this not just because they don't want to but also because they lack the capability to do so. After all, western democracies, including the United States, have proved remarkably reluctant to go to war without allies. Given this overwhelming preference for coalition warfare, it seems absurd to continue to develop strategy in narrowly national terms.

So the first source of our current ills is that we have had to put the conduct of war, as opposed to the avoidance of war, back into our thinking. The second is the confusion as to whether Western democratic states are actually at war. Since 9/11 their national leaders have found it increasingly hard not to respond to terrorist threats by suggesting that they are: both David Cameron and François Hollande were cases in point, as of course was George W. Bush. National leaders have used the rhetoric of national mobilisation, so evoking the memory of the Second World War or – for France – the wars of the Revolution. Their peoples seem increasingly not to agree. They see no direct evidence of the fact: they are not conscripted, their taxes are not increased to pay for larger armed forces, and their daily lives continue to be conducted according to the routines of peacetime. True, their (by and large professional) armed forces are deployed in war zones and suffer casualties. And yet the west persists (and this is the third source of its confusion) in calling the conflicts in which its states have been engaged 'wars of choice'. Most citizens of Western democracies probably accept that there are occasions when it may be necessary to resort to war, especially when they put necessity in the context of the Second World War, but they find it hard to regard war as a something they opt to do, as opposed to it being forced on them. Choice in this context suggests frivolity in the decision to embark on an enterprise with momentous and destructive implications.

Finally, we have confused strategy in theory with strategy in practice. Those who thought about war in the wake of Napoleon were well aware of the importance of this distinction, none more so than Ferdinand Foch, the allied generalissimo in 1918. In 1914 Foch had served for 40 years and reached corps command without having seen action: his views on war were shaped by study and reading, and were articulated before the First World War in the lectures he had delivered at the Ecole de Guerre. He was a strategic theorist *par excellence*. He then fought to defend his country for 4 years, at the Marne, around Ypres and on the Somme. By 1918 he realised that, important though study was, it was not sufficient: that, although strategy

> may be simply understood after it has been practised, it is not a simple thing to put into practice. What is required is the ability, in special circumstances, to appreciate the situation as it exists, shrouded in the midst of the unknown.[14]

Theory, and the plans which flow from it, were only preparatory and preliminary. 'Plans must be adapted to circumstances', he said after the First World War when reflecting on the causes of the German defeat. The capacity to respond to the immediate situation determined success: 'The secret of Napoleon was to meet events half-way so that he could control them, instead of waiting and allowing them to over-ride him'.[15] During the Cold

---

[14]Commandant Bugnet, *Foch Talks*, translated by Russell Green (London: V. Gollancz Ltd 1929), 191.
[15]Raymond Recouly, *Marshal Foch: His Own Words on Many Subjects*, translated by Joyce Davis (London: Thornton Butterworth 1929), 100, 128.

War, because it never turned hot, plans were not trumped by circumstances, and so theory prevailed, untrammelled by contact with reality.

Strategic theory, as developed before 1914 by professional soldiers like Foch and his predecessors, sought to understand war as a general phenomenon. It did so in two ways. One, the doctrinal thread, owed its intellectual origins to Jomini, whose biography by Jean-Jacques Langendorf is revealingly and rightly called *Faire la guerre*, how to 'do' war.[16] Jomini's method centred on planning, and laid down the principles which would deliver strategic success, using cartography and geographical awareness to manoeuvre and to master lines of communications. The other approach, which we owe to Clausewitz, is best understood in the title of Raymond Aron's book on him, *Penser la guerre*, or how to 'think' about war.[17] Here the role of theory is to prompt strategic thought as a route to comprehension rather than to action. The departure point is the need to think before acting, recognising that it is important to get the questions right before jumping to conclusions. If the initial questions are wrong, then the answers are likely to be too.

Strategic theory, certainly until 1914 and even until 1945, sought continuities. It was a dialogue between the present and the past, aiming to put current conflicts in context by comparing them with those of history. So both Clausewitz and Jomini turned to the wars of Frederick the Great to provide a way of measuring their own experiences in the wars of the French Revolution and Napoleon. History enabled them to understand what was enduring and what novel, and to distinguish between the two – in other words, to manage the relationship between continuity and change. Once it had sorted out the new from the familiar, theory then had to decide whether the innovations were lasting or temporary, and if the former to incorporate them in the body of received strategic wisdom. Strategic theory constructed on these lines would be slow to anticipate change and indeed might be slow to recognise it. In an area of study whose prime motivation is prudential and anticipatory, that is a significant disadvantage. Since 1945 strategic theory has used disciplines other than history, such as game theory, mathematics and economics, to give it a better purchase on the future, albeit without any more obvious signs of success. In the process, its hold on historical precedent has become tenuous, and so it has not been as effective in its capacity to recognise and assimilate what is genuinely new as opposed to what only seems to be new. This is why Colin Gray and others have used history in the ways which they have – to reinstate the perceptions derived from the continuities in strategic theory.

---

[16]Jean-Jacques Langendorf, *Faire la guerre: Antoine-Henri Jomini* (2 vols, Geneva: Georg 2001–2004).
[17]Raymond Aron, *Penser la guerre, Clausewitz* (2 vols, Paris: Gallimard 1976).

Strategy in practice is not like that. This is so for three reasons. The first and most important is that there is no universal character to war. War may have its own nature: it rests on reciprocity, on the clash of wills with which Clausewitz begins chapter 1, book I of *Vom Kriege*. Violence, the business of killing and being killed, lies at its core. It requires courage of its participants and it is shaped in circumstances which are confusing and profoundly challenging. But what follows from all this is that each war is in practice very different, possessed of its own characteristics. War at sea is not the same as war on land, and within war, at sea, the battle of Salamis was very different from the battle of Jutland, and both were different again from the battle of Midway. Recent military experience makes a similar point. For those who served in both Iraq and Afghanistan, there were some similarities between the two theatres. Ultimately the United States and its allies tried to apply common principles in the shape of counter-insurgency warfare. In practice, however, the two countries provided operating environments which were fundamentally different, in culture, history, geography, religious practice, literacy and economic potential. All these could make the ready acceptance of superficial commonalities positively dangerous. Generalised operating principles make sense for hierarchical organisations like armies, but they militate against the exploration of difference, even if they do not absolutely preclude it. This is why we study military history: not because wars are the same, but to understand how they differ. History teaches cause and effect before it says anything about generalizable and transferable principles. Marc Bloch, the French medieval economic historian who served his country as a soldier in 1914 and 1940, made this point when he addressed the problem of why the French army, which had emerged victorious in the First World War, had been so ignominiously defeated in 1940: the value in history is that it 'is, in its essentials, the science of change'. The French army of 1940 had not realised this, believing instead that history repeated itself, a point which every historian knows to be untrue.[18]

The second challenge for strategy in practice is that we are told that wars are the continuation of politics by other means, when they are not, and particularly not for democratic states. Policy generates change over very short lead-times. The 1997–1998 British Strategic Defence Review aspired to be long-term and prudential: it was even praised because it was 'strategy-led'. However, its assumptions were overthrown by the responses of the United States and its allies to the 9/11 attacks. The review had set out to create a maritime-air capability for expeditionary war, which assumed that short-term overseas interventions would be the norm. In practice Britain

---

[18]Marc Bloch, *Strange Defeat: A Statement of Evidence Written in 1940*, translated by Gerard Hopkins (New York: Norton 1999), 117–118. Bloch's portrayal of French military education in the inter-war period overplays the influence of Napoleon and underplays the real effort to engage with the lessons of the First World War, but his more general argument stands.

fought two protracted campaigns in Iraq and Afghanistan in which the army led, and in which the capability decisions of 1997–1998, predicated on aircraft carriers and multi-role combat aircraft, were trumped by so-called 'urgent operational requirements', running from armoured vehicles to drones. For most western democracies, going to war represents a decision for change, not continuity. It reflects a recognition that the previous policy has failed and that a new approach is required.

Furthermore, recent conflicts show that it is not just the initial decision to go to war which resets policy. The destabilising influences of democratic politics permeate war once hostilities have commenced, often undermining strategy itself. The effort to establish a coherent campaign plan for the war in Afghanistan was constantly reset by the need to adjust to changing political objectives. For some the aim was the defeat of al-Qaeda in Afghanistan; for others, it was the overthrow of the Taliban; and for a third group, it was the construction of a viable Afghan state capable of running its own affairs in ways that respected human rights and the rule of law. The United States itself oscillated along this spectrum, and its allies positioned themselves on it at different points at different times. The effects, while preventing strategy from developing any consistency, at least proved – albeit negatively – that war can indeed be the continuation of policy by other means. Governments, however, run the risk that policy itself will be subordinated to the contingencies of war, particularly if the war becomes protracted and then goes in unpredicted directions. They too easily forget that the enemy has a vote and is ready to meet force with force. Loss of life in war all too often becomes a reason for continuing it, in order to justify and hallow the loss: 'our boys' cannot have died in vain. The effect of killing and death can render a return to the *status quo ante* all but impossible. That is true for pragmatic reasons as well as emotional. In August 1914 the British Parliament rushed through the Defence of the Realm Act in order to deal with German spies. The orders passed under the terms of the act left a legacy after the end of the war in 1918 which continued to shape the patterns of daily life in Britain for decades, from the opening hours of pubs to the introduction of British summer time. In 2009, the United States formally abandoned the global war on terror when its progenitor, George W. Bush, was succeeded as president by Barack Obama, but many of the structures put in place then – ranging from airport security to the detention facility in Guantanamo bay – remain.

Thirdly, once war becomes policy, politicians in democratic states become the *de facto* strategists. Few of them, however, have studied strategic theory. This was General Ripper's point: strategy is studied in staff colleges and military academies by those whose careers will be in uniform. Between 1864 and 1870 the Prussian minister-president, Otto von Bismarck, used war to further the interests of his policies better than most in the history of modern Europe. He unified Germany by 1871 through a sequence of wars, which were short enough to be concluded before other powers were drawn in and sufficiently decisive to ensure that they did not in fact trump his policy, but served it. Bismarck was the exception who was used to prove the rule, but he did not consider himself a strategist. Today we associate the trite generalisation that war is the continuation of policy by other means with Carl von Clausewitz, another Prussian and one whose study of war, *Vom Kriege*, first published just over three decades before, had come out in a revised edition as recently as 1853. However, there is no evidence that Bismarck ever read Clausewitz. Bismarck owed his success to common sense, not to strategic theory.

As a statesman not formally versed in strategic theory, he was then – and remains today – hardly unusual. As George W. Bush girded the United States for war in 2002, Eliot Cohen published an influential study of democratic political leaders as strategic practitioners. Called *Supreme Command*, it looked at four case studies – Abraham Lincoln in the American Civil War, Georges Clemenceau in the First World War, Winston Churchill in the Second World War, and David Ben-Gurion in Israel's War of Independence. Each of them read widely, but only Churchill had read deeply in strategic theory: as a young man, for all that he was a cavalry subaltern who liked to trumpet his lack of formal education, he asked his doting mother to send him the latest professional publications. This was an era, sandwiched between the Franco-Prussian War and the First World War, when military publishing flourished, and did so internationally. But Churchill did not overtly refer to those texts as a wartime leader in either world war. Instead, his early morning reading as prime minister was dominated by the latest intelligence intercepts. For him, as for Cohen's other three subjects, strategy was a profoundly pragmatic business, shaped by the realities of the moment, as they endeavoured to meet contingency head on while retaining an overall sense of direction.

We might question how far they provide a model of supreme command for the wars of the early twenty-first century. Because they were engaged in wars of national survival, their premierships were defined by their mastery of strategy. In this, they differed from George W. Bush and even more from his peers among the United States's allies. For them the 'global war on terror', despite the grandiloquence of its title, never subordinated the other affairs of state to the needs of war. Policies pursued in relation to housing or health or education continued to have their own priorities and values, and demanded their attention at least as much as did foreign affairs and military operations. The war was not used to raise taxes or to explain the fiscal deficit or to justify the financial crash of 2008. By contrast, for Cohen's gang of four, the policy was war, and their wars dominated their policies in other areas, including the management of the economy. The practitioners of strategy in the early twenty-first century have never allowed themselves – or been allowed – to develop that single-minded focus, even if they have had the appetite for it.

Those versed in maritime strategy might with good reason argue that this account exaggerates the polarity between strategy in practice and strategic theory, and does so because it over-estimates the role of continental Europe, with its land wars waged by armies, and underplays the role of empire, trade and maritime expansion. Wars at sea have struggled to find their place in mainstream strategic theory. The early running in the post-Enlightenment tradition of writing on war was made by men like Jomini and Clausewitz, soldiers with a theoretical caste of mind, and focused on land warfare. Writers like Alfred Thayer Mahan and Julian Corbett did not enter the fray until the turn of the nineteenth and twentieth centuries, almost a hundred years later. Both of them wrote books which were directly informed by their landed predecessors, Mahan's by Jomini and Corbett's by Clausewitz, and yet both had a more ambivalent relationship with the centrality of battle in warfare. As Corbett put it, when addressing the 'maxim' that the only way of securing command of the sea 'is to obtain a decision by battle against the enemy's fleet', 'nothing is so dangerous in the study of war as to permit maxims to become a substitute for judgment'.[19] He went on to stress the role of blockade rather than battle in securing command of the sea. Furthermore, both he and Mahan were well aware that the foundations of maritime power were built in peace rather than in war. Maritime power in 'a globalised world', a phrase that seemed as appropriate to them as it does today, with its patterns of international trade sustained by the City of London and the convertibility of the pound sterling, depended on the acquisition of bases and the long-term construction of fleets. For imperial and trading powers like the United States and Britain, maritime strategy related much more directly to the pursuit of policy outside war than did the understanding of strategy pro-moted by Jomini or Clausewitz. It was also a persistent feature of national policy, not – as major war was – an episodic and infrequent response to abnormal circumstances.

Imperial and maritime power was also predicated on a 'rules-based international order' – or the capacity to challenge those rules. If maritime strategy possessed earlier origins than those associated with Mahan and Corbett, or even than Jomini and Clausewitz, they are to be found in the development of international law. The career of Hugo Grotius, the author *De jure belli ac pacis*, published in 1625, highlights two important points about

---

[19]Julian Corbett, *Some Principles of Maritime Strategy* (1911), edited by Eric Grove (Annapolis, MD: Naval Institute Press 1988), 167. Lukas Milevski, *The Evolution of Modern Grand Strategic Thought* (Oxford: Oxford University Press 2016) sees the origins of its subject in maritime strategy.

the effect of the law on strategic theory. First, the sea provided the test bed for his ideas on the relationship between law and war before Europe was assailed by the impact of the Thirty Years War (1618–1648). Grotius was employed by the Dutch East India Company to represent it in a case concerning prize law at sea. He produced a commentary on the law of prize and booty, and the part which addressed the freedom of the seas (and asserted it as a principle) was published in 1609. As the power of the centralised state extended in terms of territorial control and effective administration, particularly in Europe after 1648, the sea – by contrast – remained an unregulated space, where piracy and brigandage co-existed with more regular forms of war. The practice of state privateering – giving vessels letters of marque to attack enemy cargos – was not finally outlawed until the treaty of Paris at the end of the Crimean War in 1856. It was a fragile equilibrium: in the First World War international maritime law was challenged in theory and in practice by both sides. After it, Britain remained as resistant to the principle of the freedom of the seas (proposed by the United States as part of the peace settlement), as it had been to Grotius 300 years earlier.[20]

Corbett, himself trained as a lawyer, was clear that economic warfare was a cardinal weapon in Britain's armoury, a point which he had developed in *Some Principles of Maritime Strategy* before the First World War, and which he and others believed had been vindicated by its outcome. International law – and specifically the Declaration of London of 1909 – tried to pin down a tight definition of contraband, in order to ensure that belligerent states could only seize goods which were directly defined as munitions of war. This was not how Corbett saw the purposes of blockade. 'If on land you allow contributions and requisitions, if you permit the occupation of towns, ports, and inland communications, without which no conquest is complete and no effective war possible, why should you refuse similar procedure at sea where it causes far less individual suffering?', he argued in 1911. What really brings war to an end, he went on, is the exertion of 'pressure on the citizens and their collective life'.[21]

Corbett's version of economic war, and Britain's practice of it in the First World War, legitimised attacks on civilians. It was the revolution in Germany, not its army's defeat on the western front, which had ended the war in November 1918, in the eyes not just of German generals anxious to argue that they had been 'stabbed in the back' but also of maritime strategists. This was the second contribution of maritime strategy to strategic theory:

---

[20]John W. Coogan, *The End of Neutrality: The United States, Britain and Maritime Rights, 1899–1915* (Ithaca, NY: Cornell University Press 1981); Isabel Hull, *A Scrap of Paper: Making and Breaking International Law during the Great War* (Ithaca, NY: Cornell University Press 2014); Bernard Semmel, *Liberalism and Naval Strategy: Ideology, Interest, and Sea Power during the Pax Britannica* (Boston: Allen & Unwin 1986).
[21]Corbett, *Some Principles of Maritime Strategy*, 97.

that, especially in 'total wars' fought by democracies, the people of the nation had a responsible role in the making of strategy and in the making of war. Economic warfare could target the people, especially by denying them food, and so force them into revolution against the state which had taken them into war. The apparent success of this approach fed the arguments for strategic bombing almost from the outset. As Giulio Douhet looked at the First World War from the perspective of Italian neutrality in 1914, he anticipated it would end in revolution, and by 1921 this belief underpinned his arguments for the efficacy of air power.[22] In 1944 the western allies believed that Germany would collapse before the end of that year because its people, under sustained attack from the air, would overthrow Hitler, just as in 2003 the Americans convinced themselves that Iraqis would welcome them because they would have overthrown Saddam Hussein.[23]

Here were approaches to strategy, generated in peacetime, linked to the growth of democracy, and bridging the gap between the use of war and its utility for national policy, which took strategy in new directions. In the United Kingdom, the phrase 'grand strategy' gained currency in the 1920s and 1930s, and it was employed successfully as an organising tool in the Second World War. It turned the concepts of full national mobilisation, coalition warfare and multi-front campaigning, waged in the three dimensions of land, sea and air, into a workable framework. After the war was over, the United States encapsulated a similar set of ideas in 'national strategy', a title which reflected the National Security Act of 1947. Congress said that the role of the National Security Council, established under the act, was to 'advise the President with respect to the integration of domestic, foreign, and military policies'.[24]

The Cold War and its accompanying vocabulary of deterrence allowed the nostrums of grand and national strategies to continue to hold good. The 'hot wars' waged within its framework were not seen as wars of national survival for the major powers, but were defined as limited wars, small wars, counter-insurgency campaigns and low-intensity operations. They might be lumped together as part of the wider ideological struggle against communism but, as long as that bigger contest was contained, that argument was itself a theoretical as well as an ideological construct. War and strategy were set on divergent courses without anybody really noticing until after the end of the Cold War. Since then broad notions of grand strategy have struggled to sustain their momentum. The reduction of the role of deterrence in public

---

[22]Thomas Hippler, *Bombing the People: Giulio Douhet and the Foundations of Airpower Strategy, 1884–1939* (Cambridge: Cambridge University Press 2013).
[23]F. H. Hinsley, et al., *British Intelligence in the Second World War: Its Influence on Strategy and Operations* (4 vols, London: Her Majesty's Stationery Office 1979–1990), 3/2, 365; Richard Overy, *The Bombing War: Europe 1939–1945* (London: Allen Lane 2014).
[24]William Burr, 'National Security Council', in John Whiteclay Chambers II (ed.), *The Oxford Companion to American Military History* (Oxford: Oxford University Press 1999), 470.

rhetoric; the belief – however naïve – that major war has had its day; the elevation of terrorism in the public discourse on national security: all militate against broad definitions of strategy. At the same time, especially since 9/11, there have been wars in abundance but none which ostensibly merits inclusion within a big concept like grand strategy. To revalidate itself, grand strategy – at least in the US – has assumed more open-ended and long-term objectives, concerned as much with domestic arguments about national strength and resilience, including health, education and the institutions of democracy. The specifics of individual wars have not readily accommodated themselves to this narrative, which in any case tends to downplay contingency and even the impact of external adversaries.[25]

Outside the United States, its allies are less sure that they have any need for grand strategy at all, not least because of their subordinate status, and the Americans reinforce this conviction by insisting that only a great power has need of a grand strategy. For Britain specifically, grand strategy has been dismissed as a legacy of empire.[26] The most obvious symptom of this malaise relates to the poverty of the public debate around one of the principal props – and even origins – of grand strategy. Although the debate on strategy was re-energised by the effects of the global war on terror, and of the wars in Iraq and Afghanistan, that on maritime strategy specifically has stuttered.[27] The naval professionals of traditional sea powers, including the United States and the United Kingdom, complain about the sea blindness of their peoples. The latter, they say, fail to appreciate their dependence on the security of the sea, and don't notice the centrality of the sea to world trade, the inter-dependence of national economies in an era of globalisation, and the proximity of most of the world's population to the sea. None of this is untrue or unimportant, but it is also a statement of the obvious, which does not in itself explain why powers need highly sophisticated aircraft carriers, destroyers, frigates or nuclear submarines, all optimised for conflicts between peer rivals. It does not draw the specific connection of navies to war, to what classical maritime strategy called sea control or sea denial. Instead, it makes

---

[25]Hal Brands, *What Good is American Grand Strategy? Power and Purpose in American Statecraft from Harry S. Truman to George W. Bush* (Ithaca, NY: Cornell University Press 2014). Ionut Popescu, *Emergent Strategy and Grand Strategy: How American Presidents Succeed in Foreign Policy* (Baltimore: Johns Hopkins University Press 2017) makes the case for responding to contingency in grand strategy.

[26]House of Commons Public Administration Committee, *Who does UK National Strategy?* First report of session 2010–11, HC435 (18 October 2010), 8–9; Williamson Murray, 'Thoughts on Grand Strategy', in Williamson Murray, Richard H. Sinnreich and James Lacey (eds.), *The Shaping of Grand Strategy: Policy, Diplomacy, and War* (Cambridge: Cambridge University Press 2011), 1.

[27]Exceptions to this generalisation include Geoffrey Till, *Seapower: A Guide for the 21st Century* (Abingdon: Routledge 2013), Chris Parry, *Super Highway: Sea Power in the 21st Century* (London: Elliott and Thompson 2014), and Daniel Moran and James Russell (eds.), *Maritime Strategy and Global Order: Markets, Resources, Security* (Washington DC: Georgetown University Press 2016).

a virtue of the diplomatic and political leverage provided by a maritime presence. In so doing it links the theory of sea power to the practice of policy and commerce, and not to strategy itself.

The response of soldiers to the evidence of a divergence between strategy in theory and strategy in practice since 9/11 has been very different. They have reverted to the position adopted by their nineteenth-century predecessors. Traditional strategic theory responded to the unpredictable effects of policy on the conduct of war by effectively discounting it. By seeing strategy as a self-contained area of military professional competence, they stressed its relationship to tactics, not to politics, to the use of the battle for the purposes of the war, not to the use of war for the purposes of policy.

This way out was foreshadowed before the end of the Cold War by the emergence of operational art in military thought in the 1980s. Two pressures contributed to this. One was the defeat of the United States in Vietnam. Operational doctrine became a tool to rebuild the army's sense of self-regard, best embodied in *Field Manual 100–5: Operations*, in its 1982 edition. The other was the need for NATO armies on the inner German border, pre-eminently those of Britain and West Germany itself, to address how they would deal with a conventional attack launched by the Warsaw Pact. NATO, under pressure not to escalate to nuclear weapons too soon, if at all, had to come up with more effective conventional solutions to the conduct of land warfare. They were not strategy: that was set by the stand-off in Europe between NATO and the Warsaw Pact. But they looked like strategy as defined by military thinkers from Clausewitz to Foch, and they presented policy-makers with operational alternatives which could shape their strategies in practice. They also re-empowered soldiers in the making of strategy, by opening the door to military expertise and revalidating old principles like manoeuvre and surprise. The approach was never tested in Europe, but the allies' success in the first Gulf War of 1990–91, when they were able to defeat a numerically stronger Iraqi army fighting a defensive battle, provided an operational solution to a seemingly intractable political problem. The 1992 memoirs of the allied and US commander in the Gulf War, Norman H. Schwarzkopf, are indicative. The index's heading for strategy said 'see military art', and the sub-headings under military art ran from 'breaching operations' to 'unity of command' by way of such subjects as 'desert warfare, training for' and 'logistics'. Schwarzkopf did not address strategy as Steven Jermy or John Stone was to do in 2011: he eschewed politics and his own index entry described himself as a 'military strategist and tactician'. As an infantry captain at Fort Benning, Schwarzkopf had won the George C. Marshall Award for Excellence in Military Writing with an essay which described a day of fighting. Only at the end was the protagonist revealed as

Julius Caesar. Schwarzkopf said he had written it 'to demonstrate the time-lessness of the principles of war'.[28]

When confronted with shifting policies in relation to the wars in Iraq and Afghanistan, generals who had been subalterns in the 1980s reverted to what for them provided continuity and operational direction. In 2008 General James Mattis, the US Joint Force Commander and a Marine, wrote of the need 'to return to time-honored principles and terminology that our forces have tested in the crucible of battle and that are well grounded in the theory and nature of war'.[29] Others in the British and French armies made similar points at the same time. These generals were using strategy in theory to put shape on the wars that they faced, and did so by generalising – as Jomini and Clausewitz had done – from the experience of one war by putting it in the context of war as a general phenomenon. It was an important corrective. It put war back into strategy; it served to remind politicians that the enemy would frustrate their plans; that wars have their own dynamics; and that war is not an instrument to fulfil the ends of policy in a pure and directed way. In short, war is a reactive process, which itself can change policy.

As in earlier efforts to do that, military thought was more comfortable with the relationship between strategy and tactics than that between strategy and policy. Three interconnecting problems therefore persisted. First, this approach did not of itself reconnect strategy in theory with strategy in practice. Secondly, in their focus on war soldiers were saying things which did not – despite the era of joint warfare – relate to what sailors were saying about strategy, which was more about trade and diplomacy than about fighting. The two were speaking past each other. Thirdly, the soldiers' operational focus was not tied into defence policies, with the obvious danger that states generated national security strategies which failed to learn lessons from the immediate past. The most obvious example is the enquiry into the war in Iraq chaired by Sir John Chilcot for the British government, which reported in 2016, too late to influence the 2010 or 2015 National Security Strategies, and with little public expectation that it would shape their successors.

As a result, the United States and its allies have persisted in strategic failure. Because operational thought has been more coherent than strategy, the former has entered the space vacated by the latter, posing as something it is not, and failing to deliver as a result.[30] There are sins of omission and

---

[28]H. Norman Schwarzkopf, *The Autobiography: It Doesn't Take a Hero*, written with Peter Petrie (London: Bantam Books 1992), 97, 145, 521–522, 526, 527.

[29]James Mattis, 'USJFCOM Commander's Guidance for Effects-Based Operations', *Parameters* (Fall 2008), 18–25, 18.

[30]Two of the most forceful critics were Douglas Porch, *Counterinsurgency: Exposing the Myths of the New Way of War* (Cambridge: Cambridge University Press 2013), and Gian Gentile, *Wrong Turn: America's Deadly Embrace of Counter-Insurgency* (New York: The New Press 2013).

commission on both sides of the military-civil divide. Soldiers have pre-sented counter-insurgency doctrine as a strategy when in fact it is an operational method. Politicians have become embroiled in tactical fixes to strategical problems, most openly through the use of drones (or unmanned aerial or remotely piloted vehicles) for the targeted killing of enemy leaders, and more discreetly through the deployment of special forces. These tactical and operational methods are exactly that: means which need to be allied to ends if they are to have strategic purchase.

Solutions to these challenges are not straightforward, but they need to begin with an awareness of the distinctions between operations and strat-egy, and between 'military' strategy and 'grand' strategy, however, confused those differences become in practice and even in planning. Above all, debate should be informed by the recognition of the distinction between strategy in theory and strategy in practice. That does not mean that these two approaches are choices: they are neither alternatives nor optional. Both are required, and they need to be related one to the other. Strategy in theory, knowledge of war's nature, has to inform strategy in practice. The former seeks continuity and ongoing principles; the latter embraces con-stant change and unpredictability. Each has a role in helping the other.

The intellectual framework for this relationship requires institutional expression. Coherent governmental machinery can give effect to these aspirations, shaping the dialogue between the two. Theory enables an informed discussion between the chiefs of staff and their political masters, in order to find a viable balance between military means and political ends. As Corbett wrote in 1911,

> Nor is it only for the sake of mental solidarity between a chief and his subordinates that theory is indispensable. It is of still higher value for produ-cing a similar solidarity between him and his superiors at the Council table at home. How often have officers dumbly acquiesced in ill-advised operations simply for lack of the mental power and verbal apparatus to convince an impatient Minister where the errors of his plan lay? [31]

He went on, 'Conference is always necessary, and for conference to succeed there must be a common vehicle of expression and a common plane of thought. It is for this essential preparation that theoretical study alone can provide, and herein lies its practical value.'[32]

Another way of viewing the problem is to ask three questions. First, who thinks about strategy? Second, who decides strategy? Third, who does strategy? Although today academics and think tanks undoubtedly address the first question, it is professional military education which gives strategic thought the most sustained attention. It is nonsensical not to give those

---

[31]Corbett, *Some Principles of Maritime Strategy*, 5.
[32]Corbett, *Some Principles of Maritime Strategy*, 8.

who have been the beneficiaries of that schooling a voice in council. However, the answer to the second question – at least in a modern democracy – is the government, which takes the political responsibility for the strategy which is eventually adopted. Its policy may be the right one, but that is not the end of the story. The decision requires implementation, and here the armed forces re-enter the equation, at least in the event of war. Failures in execution or successful responses by the enemy change the situation on the ground. At that point, the armed forces have to be involved in any decision about the need to moderate the ends which are being sought or about finding a different route to their fulfilment.

In these deliberations, strategy needs to be modest about itself and about what it can deliver. It is, after all, more of an art than a science, and it behoves those who think about it and those who practice it not to be too brazen about its status.[33] Its principles may be guidelines, but, as strategic theorists who are worth their salt have stressed, they are not rules. Precisely because strategy is a pragmatic business it lacks the clarity and purity which strategic theory so often seeks. Strategy has to be reactive as well as predictive, and it must be open to new evidence, whether presented as intelligence or acquired as experience. Strategic theory 'can at least determine the normal', Corbett wrote, and 'having determined the normal, we are at once in a stronger position', not least because 'we can proceed to discuss clearly the weight of the factors which prompt us to depart from the normal'.[34]

Disclosure statement

No potential conflict of interest was reported by the author.

---

[33]Everett Carl Dolman, *Pure Strategy: Power and Principle in the Space and Information Age* (London: Frank Cass 2003), 188–194, makes the case for strategy as an art.
[34]Corbett, *Some Principles of Maritime Strategy*, 9.

Bibliography

Aron, Raymond, *Penser la guerre, Clausewitz* (2 vols, Paris: Gallimard 1976).

Bloch, Marc, *Strange Defeat: A Statement of Evidence Written in 1940*, translated by Gerard Hopkins (New York: Norton 1999).

Brands, Hal, *What Good Is American Grand Strategy? Power and Purpose in American Statecraft from Harry S. Truman to George W. Bush* (Ithaca, NY: Cornell University Press 2014).

Bugnet, Commandant, *Foch Talks*, translated by Russell Green (London: V. Gollancz Ltd 1929).

Burr, William, 'National Security Council', in John Whiteclay Chambers II (ed.), *The Oxford Companion to American Military History* (Oxford: Oxford University Press 1999).

Clausewitz, Carl von, *On War*, translated and edited by Michael Howard and Peter Paret (Princeton: Princeton University Press 1976).

Coogan, John W., *The End of Neutrality: The United States, Britain and Maritime Rights, 1899–1915* (Ithaca, NY: Cornell University Press 1981).

Corbett, Julian, *Some Principles of Maritime Strategy 1911*, edited by Eric Grove (Annapolis: Naval Institute Press 1988).

Coutau-Bégarie, Hervé, *Traité de stratégie* (6th, Paris: Economica 2008).

Dolman, Everett Carl, *Pure Strategy: Power and Principle in the Space and Information Age* (London: Frank Cass 2003).

Gaddis, John Lewis, *On Grand Strategy* (London: Allen Lane 2018).

Gentile, Gian, *Wrong Turn: America's Deadly Embrace of Counter-Insurgency* (New York: The New Press 2013).

Gray, Colin S., *Modern Strategy* (Oxford: Oxford University Press 1999).

Gray, Colin S., *The Strategy Bridge: Theory for Practice* (Oxford: Oxford University Press 2010).

Hart, Liddell, *Basil, the Decisive Wars of History: A Study in Strategy* (London: G. Bell & Sons Ltd 1929).

Hart, Liddell, *Basil, When Britain Goes to War: Adaptability and Mobility* (London: Faber & Faber 1932).

Hart, Liddell, *Basil, the Strategy of the Indirect Approach* (London: Faber & Faber 1942).

Hinsley, F. H., et al., *British Intelligence in the Second World War: Its Influence on Strategy and Operations* (4 vols, London: Her Majesty's Stationery Offices 1979–1990).

Hippler, Thomas, *Bombing the People: Giulio Douhet and the Foundations of Airpower Strategy, 1884–1939* (Cambridge: Cambridge University Press 2013).

House of Commons Public Administration Committee, *Who Does UK National Strategy?* First report of session 2010-11, HC435, (18 October 2010).

Howard, Michael, *The Causes of Wars and Other Essays* (London: Temple Smith 1983).

Hull, Isabel, *A Scrap of Paper: Making and Breaking International Law during the Great War* (Ithaca, NY: Cornell University Press 2014).

Jermy, Steven, *Strategy for Action: Using Force Wisely in the 21st Century* (London: Knightstone Publishing 2011).

Joint Doctrine Publication 01: Joint Operations (Joint Doctrine and Concepts Centre March 2004).

Langendorf, Jean-Jacques, *Faire La Guerre: Antoine-Henri Jomini* (2 vols, Geneva: Georg 2001–2004)

Luttwak, Edward N., *The Grand Strategy of the Roman Empire: From the First Century A. D. To the Third* (first published 1976, London: Weidenfeld & Nicolson 1999)

Luttwak, Edward N., *The Grand Strategy of the Byzantine Empire* (Cambridge MA: The Belknap Press of Harvard University Press 2009).

Mattis, James, 'USJFCOM Commander's Guidance for Effects-Based Operations', *Parameters* (Autumn 2008), 38, 18–25.

Milevski, Lukas, *The Evolution of Modern Grand Strategic Thought* (Oxford: Oxford University Press 2016).

Moran, Daniel and James Russell, (eds.), *Maritime Strategy and Global Order: Markets, Resources, Security* (Washington DC: Georgetown University Press 2016).

Murray, Williamson, 'Thoughts on Grand Strategy', in Williamson Murray, Richard H. Sinnreich, and James Lacey (eds.), *The Shaping of Grand Strategy: Policy, Diplomacy, and War* (Cambridge: Cambridge University Press 2011), 1–33.

Napoléon, *De la guerre, Presenté et annoté par Bruno Colson* (Paris: Perrin 2011).

Norman, Schwarzkopf, H., *The Autobiography: It Doesn't Take a Hero* (written with Peter Petrie, London: Bantam Books 1992)

Overy, Richard, *The Bombing War: Europe 1939–1945* (London: Allen Lane 2014).

Parker, Geoffrey, *The Grand Strategy of Philip II* (New Haven, CT: Yale University Press 1998).

Parry, Chris, *Super Highway: Sea Power in the 21st Century* (London: Elliott and Thompson 2014).

Popescu, Ionut, *Emergent Strategy and Grand Strategy: How American Presidents Succeed in Foreign Policy* (Baltimore MD,: Johns Hopkins University Press 2017).

Porch, Douglas, *Counterinsurgency: Exposing the Myths of the New Way of War* (Cambridge: Cambridge University Press 2013).

Recouly, Raymond, *Marshal Foch: His Own Words on Many Subjects* (translated by Joyce Davis, London: Thornton Butterworth 1929)

Semmel, Bernard, *Liberalism and Naval Strategy: Ideology, Interest, and Sea Power during the Pax Britannica* (Boston: Allen & Unwin 1986).

Stone, John, *Military Strategy: The Politics and Technique of War* (London: Continuum 2011).

Till, Geoffrey, *Seapower: A Guide for the Twenty-First Century* (Abingdon: Routledge 2013).

von Bernhardi, Friedrich, *Vom Kriege der Zukunft. Nach der Erfahrungen des Weltkrieges* (Berlin: E.S. Mittler & Sohn 1920).

Military strategy and peacekeeping: An unholy alliance?

Kersti Larsdotter ⓘD

ABSTRACT
Despite the increased use of military force in peacekeeping operations in the twenty-first century, these operations are not included in traditional strategic theory. In this article, I outline the logic of four strategies for peacekeeping operations – defence, deterrence, compellence and offence – and trace the use of these strategies in two consecutive UN operations in the Democratic Republic of the Congo: MONUC and MONUSCO. The article concludes that all four strategies are indeed used in the two operations, but they are neither comprehensive nor proactive, leaving the true potential of military strategy unrealised.

Strategic theory has traditionally been occupied with the use of force by state actors against various enemies. The last 20 years, however, we have seen the deployment of larger and more robust peacekeeping operations in situations where there is no peace to keep.[1] Although the peacekeeping forces are not combatants of an ongoing war, but rather deployed as a third party, guided by the peacekeeping principles of neutrality and impartiality, they are deployed in hostile situations resembling war. While several scholars question the use of force by peacekeepers,[2] I argue that these developments call for the use of force in peacekeeping to be scrutinised in the same way as the use of force in other kinds of military operations.[3]

---

[1] There are some exceptions, for example, the UN operation in DR Congo in the 1960s.

[2] Charles T. Hunt, 'All Necessary Means to What Ends? The Unintended Consequences of the "Robust Turn" in UN Peace Operations', *International Peacekeeping* 24/1 (2017), 108–31; John Karlsrud, 'The UN at War: Examining the Consequences of Peace-Enforcement Mandates for the UN Peacekeeping Operations in the CAR, the DRC and Mali', *Third World Quarterly* 36/1 (2015), 40–54; Thierry Tardy, 'A Critique of Robust Peacekeeping in Contemporary Peace Operations', *International Peacekeeping* 18/2 (2011), 152–67.

[3] Here, I follow Barry R. Posen. In his article on military responses to refugee disasters, Posen argues: 'The threat or use of force for humanitarian purposes is as much an act of strategy as is the threat or use of force to achieve geostrategic goals.' Barry R. Posen, 'Military Responses to Refugee Disasters', *International Security* 21/1 (1996), 72–111, 79.

Although traditional strategic theory is most certainly useful for analysing the use of force in peacekeeping operations, the deployment of peace-keepers as a third party changes the logic of military strategy. Instead of having 'enemies' as the main focus for a strategy, several actors need to be taken into consideration, for example, the (ex-)belligerents who have signed a peace agreement; the spoilers who have defected from the parties of the peace agreement as well as violent groups who were not included in the agreement in the first place; ex-members of the belligerents who have already disarmed; as well as the civilian population. Each of these actors has their own objectives and (military) means.

While traditional strategic theory is primarily concerned with major war or national defence, military strategy in other settings, such as limited war, counterinsurgency or humanitarian interventions, is increas-ingly included.[4] Two strands of research are of particular interest for our knowledge of military strategy in peacekeeping operations. First, there are a few studies on coercive diplomacy in peacekeeping. Ken Ohnishi, for example, analyse how coercive diplomacy was used by the International Force for East Timor.[5] While this research contributes to our knowledge of coercive diplomacy in different settings, it leaves other kinds of strategies in peacekeeping unexplored. Second, there is a growing literature exploring various military strategies in the protec-tion of civilians.[6] The protection of civilians is, however, only one of many objectives of contemporary peacekeeping, leaving us with limited knowledge about the utility of these strategies for achieving other objectives of peacekeeping operations.[7]

---

[4]On limited war and counterinsurgencies, see, for example, Lawrence Freedman, 'Ukraine and the Art of Limited War', *Survival* 56/6 (2014), 7–38; Stephen Peter Rosen, 'Vietnam and the American Theory of Limited War', *International Security* 7/2 (1982), 83–113; Hew Strachan, 'Strategy and the Limitations of War', *Survival* 50/1 (2008), 31–54; Hew Strachan, 'Strategy or Alibi? Obama, McChrystal and the Operational Level of War', *Survival* 52/5 (2010), 157–82. On humanitarian interventions, see Taylor B. Seybolt, *Humanitarian Military Intervention: The Conditions for Success and Failure* (Oxford: Oxford University Press 2007). See also, Genocide Prevention Task Force, *Preventing Genocide: A Blueprint for U.S. Policymakers* (Washington, DC: United States Holocaust Memorial Museum 2008), on genocide, and Posen, 'Military Responses to Refugee Disasters', on refugee disasters.

[5]Ken Ohnishi, 'Coercive Diplomacy and Peace Operations: Intervention in East Timor', *NIDS Journal of Defense and Security* 13 (2012), 53–77, 77. See also Alexander J. Bellamy, 'Lessons Unlearned: Why Coercive Diplomacy Failed at Rambouillet', *International Peacekeeping* 7/2 (2000), 95–114.

[6]See, for example, Arthur J. Boutellis, 'From Crisis to Reform: Peacekeeping Strategies for the Protection of Civilians in the Democratic Republic of the Congo', *Stability: International Journal of Security and Development* 2/3 (2013), 1–11; Stian Kjeksrud, 'The Utility of Force for Protecting Civilians', in Haidi Willmot, Ralph Mamiya, Scott Sheeran and Marc Weller (eds.), *Protection of Civilians* (Oxford: Oxford University Press 2016), 329–49; Paul D. Williams, *Enhancing Civilian Protection in Peace Operations: Insights from Africa* (Washington, DC: Africa Center for Strategic Studies 2010).

[7]There are some notable exceptions, for example, Hakan Edstrom & Dennis Gyllensporre, *Political Aspirations and Perils of Security: Unpacking the Military Strategy of the United Nations* (Houndmills: Palgrave Macmillan 2013); Trevor Findlay, *The Use of Force in UN Peace Operations* (Oxford: Oxford University Press 2002); John Hillen, *Blue Helmets: The Strategy of UN Military Operations* (London: Brassey's 1998).

The aim of this article is therefore to take a more comprehensive approach and to elaborate on the logic of four different strategies for peacekeeping operations, using traditional strategic theory as a point of departure. To better understand the utility of military strategy in peace-keeping operations, this article will also scrutinise the military strategy of two subsequent peacekeeping operations in the Democratic Republic of the Congo – the United Nations Organization Mission in the Democratic Republic of the Congo (MONUC), 1999–2010, and the United Nations Organization Stabilization Mission in the Democratic Republic of the Congo (MONUSCO), which is still ongoing.

The paper continues as follows. First, I will present the logic of the four main military strategies found in traditional strategic theory. I will also discuss some methodological considerations. Second, I will discuss how the logic of the four strategies plays out in peacekeeping operations, and trace the use of the strategies in the two peacekeeping operations in the Democratic Republic of the Congo. Third, I will highlight some of the problems with the use of strategy in the two missions and finally, conclude with some suggestions for the practice of military strategy in peacekeeping.

Military strategies

Military strategy can be understood in several ways. In this article, I follow Jeffrey W. Meiser. Instead of simply understanding strategy as the relationship between ends, ways and means, I understand strategy as 'a theory of success'. This means that I include the causal explanations of how a given action will cause the achievement of the desired ends.[8] This also means that I am interested in the decision maker's understanding of how certain means and ways can achieve the desired ends or objectives.

Although most scholars seem to agree that military strategy is important, the question of how to achieve desired ends by using force is extensively debated. One of the main distinctions in the literature is between theories where the aim is to influence the *decision* of an enemy to use force, and theories where the aim is to influence the *capacity* of an enemy to use force. In the first camp, we find strategies of compellence and deterrence. In the second camp, we find strategies of offence and defence.[9]

---

[8]Jeffrey W. Meiser, 'Ends + Ways + Means = (Bad) Strategy', *Parameters* 46/4 (2016/2017), 81–91, 86. This understanding also follows Barry Posen's understanding of grand strategy. Barry R. Posen, *The Sources of Military Doctrine: France, Britain, and Germany between the World Wars* (Ithaca, NY: Cornell University Press 1986), 13.

[9]For an overview of all four approaches, see, for example, Robert J. Art, 'To What Ends Military Power?' *International Security* 4/4 (1980), 3–35; Peter Viggo Jacobsen, *Western Use of Coercive Diplomacy: A Challenge for Theory and Practice* (Houndmills, Basingstoke, Hampshire: Palgrave Macmillan 1998); David J. Lonsdale, 'Strategy: The Challenge of Complexity', *Defence Studies* 7/1 (2007), 42–64.

According to strategies of coercion,[10] success in war is achieved by using the threat of military force, as well as limited military force, to persuade an enemy to behave in a way it would otherwise not do. Coercion is *not* about making an enemy defenceless, but persuading it not to use organised violence, despite having the capacity to do so.[11] The coercive use of force can both be active and passive in nature, i.e., compellence and deterrence.

The aim of compellence is to induce an enemy to take a certain action. It aims at changing the status quo and is therefore active in nature. The coercer can either use the threat of military force to compel an enemy to give up something of value without resistance (for example, to withdraw from territory), or to make it stop or undo an action that is already initiated (for example, stop an ongoing attack).[12] The aim of deterrence is instead to discourage an enemy from taking certain actions in the first place, and is therefore passive in nature. It seeks to preserve the status quo. It is up to the opponent to take the actions starting a war.[13] While both compellence and deterrence involve the communication of threats, compellence uses threats to coerce an enemy to take a certain action, while deterrence uses threats to persuade an enemy not to take any action.

According to strategies with the aim of influencing the capacity of an enemy, the use of force can also be active and passive in nature, i.e., offence and defence.[14] The aim of offensive strategies, or what Thomas C. Schelling termed brute force, is to defeat an enemy.[15] Compliance only takes place when an enemy no longer has the capacity to continue fighting.[16] Hence, the main difference between brute force and compellence is that brute force seeks to impose compliance on an enemy while compellence seeks to persuade an enemy to comply.[17] The aim of defensive strategies is to mitigate the effects of an attack. Defence reduces the enemy's possibility to damage the defender.[18] The main difference between defence and deterrence is that whereas defence seeks to reduce our own prospective costs

---

[10]I am using coercion in the same way as for example, Daniel Byman and Matthew Waxman, i.e. I include both deterrence and compellence. Daniel Byman and Matthew Waxman, *The Dynamics of Coercion: American Foreign Policy and the Limits of Military Might* (Cambridge: Cambridge University Press 2002). Robert A. Pape, for example, uses coercion instead of compellence. Robert A. Pape, 'Coercion and Military Strategy: Why Denial Works and Punishment doesn't', *Journal of Strategic Studies* 15/4 (1992), 423–75.

[11]Byman and Waxman, *The Dynamics of Coercion*, 3; Jakobsen, *Western Use of Coercive Diplomacy*, 11; Pape, 'Coercion and Military Strategy', 425.

[12]Jakobsen, *Western Use of Coercive Diplomacy*, 12.

[13]Colin S. Gray, *Maintaining Effective Deterrence* (Carlisle, PA: SSI, US Army War College 2003), 25; Jakobsen, *Western Use of Coercive Diplomacy*, 12; Pape, 'Coercion and Military Strategy'.

[14]However, some scholars include pre-emptive strikes in their understanding of defence. For example, Art includes both 'offensive' (first) strikes' and 'repellent (second) strikes' in his understanding of defence. Art, 'To What Ends Military Power?' 5.

[15]Thomas C. Schelling, *Arms and Influence* (New Haven, CT: Yale University Press 1966).

[16]Byman and Waxman, *The Dynamics of Coercion*, 3.

[17]For a discussion on coercion vs brute force, and different forms of coercion, see Byman and Waxman, *The Dynamics of Coercion*.

[18]Art, 'To What Ends Military Power?' 5; Glenn H. Snyder, *Deterrence and Defense: Toward a Theory of National Security* (Princeton, NJ: Princeton University Press 1961), 4.

Table 1. Military strategies.

		Objective of Military Force	
		Influence decision	Influence capability
Use of Force	Active	Compellence	Offence
	Passive	Deterrence	Defence

and risks in the event of an attack, deterrence seeks to discourage an enemy from using force by making the enemy's prospective costs and risks outweigh the gains. For an overview of the strategies, see Table 1.[19]

In traditional strategic theory, military strategy is usually understood as a *state's* use of military means to achieve its political ends, ultimately to achieve security for itself. But, before moving on to how the strategies discussed above play out in peacekeeping, some methodological clarifications are needed. First, I have chosen to equate the state with a peacekeeping operation, and not with the United Nations.[20] While the Security Council could be compared to the decision-making bodies of a state, there are several differences between various peacekeeping operations that are similar to that between various states, making this level more relevant: the actors are different, the geostrategic situation and geography are different, the military means are different, etc.

Second, since I understand strategy as a theory of success, the object of study is not the implementation of strategy, but the decision makers' understanding of how certain means and ways can achieve the desired ends or objectives of a mission. By using the level of peacekeeping operations, the relevant sources for capturing the decision makers' understanding are primarily the Security Council's resolutions – where the objectives, mandate and resources of each mission are formulated – as well as the Secretary-General's reports – where the concepts of operations are discussed. By tracing words in these documents associated with each of the four strategies discussed above, the decision makers' understanding of the use of force can be captured.

The military strategy of states and international organisations is usually classified and highly difficult to access. Although the Security Council's resolutions and the reports by the Secretary-General are not as detailed as military assessments and strategy documents, some of the deliberations of the decision makers regarding the use of force are possible to trace in these documents. The mandates reflect the overall ends and means, while the concepts of operations discussed in the Secretary-General's reports provide more details about the relationship between ends, ways and means. The reliance on resolutions and reports, instead of interviews, also makes it possible to trace the development of strategic thinking over time.

---

[19]Snyder, *Deterrence and Defense*, 4.
[20]Cf. Edstrom and Gyllensporre, *Political Aspirations and Perils of Security*; Hillen, *Blue Helmets*.

Third, while the main aim of military strategy in traditional strategic theory is to ensure the security of the state, peacekeepers are not deployed primarily to create security for the mission. I am, therefore, using the mandate as the main aim, or political end, of a peacekeeping operation. The mandate could include a variety of objectives, many of which have more to do with state-building or peace-building than security. Other objectives are clearly related to the security in the area of operations, such as the protection of civilians, the disarmament of ex-combatants and the establishment of a safe and secure environment.

Finally, the two UN missions in the DR Congo were chosen because of several reasons. The UN deployed the first operation, MONUC, in 1999, after the end of a highly violent and complex regional war. Initially, MONUC was only a small observer mission, to monitor the implementation of the Lusaka peace agreement. But, the conflict continued, and the mission was repeatedly reinforced. The subsequent mission, MONUSCO, deployed in 2010, is one of the largest UN peacekeeping operations, with almost 20 000 military troops deployed at its peak. It is, therefore, one of the cases where a military strategy is most likely to have been developed.

The mandate of the two missions has also repeatedly changed, including an increasing number of objectives. Some of the most important objectives have been the disarmament and demobilisation of foreign and Congolese armed groups, the integration of the Congolese armed forces, the holding of national and local elections, the territorial security of the DR Congo, and the protection of civilians, humanitarian personnel and United Nations personnel and facilities.[21] By including both missions in the analysis, possible changes in the military strategy can be traced over time.

Military strategies for keeping the peace

Peacekeeping operations are generally deployed after a ceasefire is established and a peace agreement is signed by the main parties of the conflict. Preserving the status quo is, therefore, the default position of peacekeeping operations, and a strategy of deterrence or defence would be the obvious choice. Traditional peacekeeping operations have a variety of means to raise the costs of non-compliance and deter the parties from deviating from a peace agreement without using military force. Military observers can, for example, monitor the belligerents' troop movements, which makes surprise attacks more difficult. They

---

[21]See, for example, UN, *S/RES/1756 (2007): Resolution 1756 (2007)* (15 May 2007). In later mandates, several of these objectives have been included as part of the protection of civilians. See, for example, UN, *S/RES/2277 (2016): Resolution 2277 (2016)* (30 March 2016), 9–10.

can also serve as a trip wire, i.e., to trigger a larger intervention if it becomes necessary.[22] But, how does this logic change when peace-keepers are deployed in environments with a high level of violence, without any peace to keep?

In the following, I will elaborate on the various logics of the use of force in peacekeeping operations according to each of the four categories discussed above; defence, deterrence, compellence and offence.[23] I will also trace the use of each strategy in the two UN operations in DR Congo.

Defence

The traditional objective of defensive military strategies is to reduce the capacity of an enemy to damage the state, either by making an enemy attack more difficult or by limiting the damages caused by enemy actions.[24] Defence is passive in nature, and the focus is on the physical capacity of the enemy. But, how does the logic of defence play out when the peacekeepers are not part of the conflict?

The most obvious example of defensive strategies in peacekeeping operations is in the defence of peacekeeping personnel and installations. This is one of the core principles of UN peacekeeping and an objective included in virtually all peacekeeping mandates. Military means are used to physically protect peacekeeping personnel and facilities when attacked by spoilers. In this way, the military observers and civilian personnel can continue to conduct their functions, contributing to a long-term solution of the conflict.[25]

Strategies of self-defence can be traced in both of the UN missions in the DR Congo. Initially, self-defence was indeed the *only* objective for the military troops. The main objective of MONUC was to observe and monitor the implementation of a ceasefire agreement. The concept of operations relied on around 500 military observers, divided into teams of four observers, dispersed throughout the country.[26] However, according to the Secretary-General, 'the levels of insecurity,

---

[22]Virginia Page Fortna, *Does Peacekeeping Work? Shaping Belligerents' Choices after Civil War* (Princeton: Princeton University Press 2008); Kersti Larsdotter, *Military Interventions in Internal Wars: The Study of Peace or the Study of War?* (Gothenburg: School of Global Studies, University of Gothenburg 2011), 58–9.

[23]This is similar to Taylor B. Seybolt's categorisation in his research on humanitarian interventions as well as Paul D. Williams categorisation in his research on civilian protection in peacekeeping operations, but including all aims of peacekeeping operations. Seybolt, *Humanitarian Military Intervention*; Williams, *Enhancing Civilian Protection in Peace Operations*.

[24]Art, 'To What Ends Military Power?' 5; Lonsdale, 'Strategy', 53; Snyder, *Deterrence and Defense*, 3–4.

[25]Sometimes, other organisations, such as NGOs and humanitarian assistance convoys are included in this kind of defensive strategies. The aim is, however, the same.

[26]UN, *S/1999/790: Report of the Secretary-General on the United Nations Preliminary Deployment in the Democratic Republic of the Congo* (15 July 1999); UN, *S/2000/30: Report of the Secretary-General on the United Nations Organization Mission in the Democratic Republic of the Congo* (17 January 2000), 12–13; UN, *S/2001/572: Eighth Report of the Secretary-General on the United Nations Organization Mission in the Democratic Republic of the Congo* (8 June 2001), 13; UN, *S/RES/1291 (2000): Resolution 1291 (2000)* (24 February 2000).

the degraded infrastructure and the difficult terrain in the country' required 'the deployment of formed units to protect military observers and civilian staff and to facilitate their activities.'[27]

Four reinforced protected infantry battalion groups, numbering a total of 3,400 troops, were proposed to deploy in the vicinity of the military observers and civilian personnel. Apart from ensuring the security and freedom of movement of UN personnel, they would also guard UN equipment and the facilities located at the sector headquarters and support bases.[28] It was clearly expressed that these forces were intended to be used in a defensive manner. In one of the mission reports from 2001, for example, the Secretary-General stressed that the function of the armed personnel 'will be exclusively to guard United Nations facilities, equipment and supplies against tampering or pilfering.'[29]

Another example of defensive strategies in peacekeeping operations is in the protection of the civilian population. Spoilers of peace agreements often target civilians as part of their strategies. This has increased the UN's focus on the protection of civilians, and several new initiatives have been developed. Lately, the protection of civilians has become one of the main objectives of peacekeeping operations.[30] The logic of defence for the protection of the population is partly different from the logic of self-defence. By physically protecting civilians, the military forces make attacks against civilians more difficult. This is similar to the logic of self-defence. But, instead of ensuring local conflict resolution efforts to continue, protecting the population is making the recruitment of spoilers more difficult at the same time as it ensures popular support for the peacekeeping mission.

There are several examples of defensive strategies in the protection of civilians in the two missions. The use of force to achieve this objective is, however, often considered in terms of humanitarian norms, i.e., that civilians should be protected from the horrors of war, rather than in strategic terms. The protection of civilians under imminent threat of physical violence became one of the objectives of MONUC already in February 2000.[31] However, it took several years before it was included in the concept of operations, and the Secretary-General repeatedly emphasised that MONUC

---

[27]UN, S/2000/30, 12. In February 2000, this was authorised by the Security Council, under a Chapter VII mandate. UN, S/RES/1291 (2000), 4. The forces were indeed also authorised with this resolution to 'protect civilians under imminent threat of physical violence', but in subsequent concepts of operations, the Secretary-General firmly stated that the troops deployed did not have the capacity to do so. See page xx for further discussion.

[28]UN, S/2000/30, 12–13; UN, S/RES/1291 (2000).

[29]UN, S/2001/128: Sixth Report of the Secretary-General on the United Nations Organization Mission in the Democratic Republic of the Congo (12 February 2001), 11.

[30]In this way, contemporary peacekeeping operations sometimes resembles humanitarian interventions, with the exception that they cannot be deployed without the consent of the belligerent parties.

[31]UN, S/RES/1291 (2000).

lacked the resources for protecting civilians. In his report to the Security Council in June 2002, for example, the Secretary-General argued that 'MONUC troops currently deployed in the Democratic Republic of the Congo are not equipped, trained or configured to intervene rapidly to assist those in need of such protection.'[32]

From late 2004, the physical defence of the civilian population gets increasingly important,[33] and in 2008, it became the first priority of MONUC. The same year, the Secretary-General reported that 'MONUC has provided protection to tens of thousands of civilians through regular patrols and to those who have sought shelter around its mobile and company operating bases across North Kivu',[34] clearly reflecting the logic of defence.

Various ways by which military means can be used to achieve the protection of civilians have been discussed and implemented in the two missions. 'MONUC-protected corridors' for the evacuation of civilians under threat, is one example.[35] Another is the designation of 'security zones'[36] and 'strategic areas'. According to the Secretary-General, these zones would be protected 'with all necessary force' through the establishment of 'a strong outer perimeter' and internal 'robust patrolling that will be reinforced by a curfew'.[37]

Deterrence

The deployment of larger military contingents in contemporary peacekeeping operations does not only make various forms of defensive strategies possible. It can also make the threat of force more credible when deterring belligerent parties to adhere to a peace agreement or a ceasefire. With a large number of forces, peacekeepers have the ability to physically punish parties that do not comply with the agreement, making the threat of force more credible. While the logic of deterrence in peacekeeping is the same as in war, i.e., to influence the decision of other actors by increasing the prospective costs of using violence, the aim is partly different. In war, the deterring actor is using the threat of force to deter attacks against itself. This is also the case in peacekeeping operations. But, the peacekeeping forces are also using the threat of force to deter attacks against the parties of

---

[32]UN, *S/2002/621: Eleventh Report of the Secretary-General on the United Nations Organization Mission in the Democratic Republic of the Congo* (5 June 2002), 11.

[33]UN, *S/2004/1034: Sixteenth Report of the Secretary-General on the United Nations Organization Mission in the Democratic Republic of the Congo* (31 December 2004), 9.

[34]UN, *S/2008/728: Fourth Special Report of the Secretary-General on the United Nations Organization Mission in the Democratic Republic of the Congo* (21 November 2008), 11.

[35]UN, *S/2009/160: Twenty-Seventh Report of the Secretary-General on the United Nations Organization Mission in the Democratic Republic of the Congo* (27 March 2009), 13.

[36]MONUSCO, 'MONUSCO Deploys to Support Security Zone in Goma – Sake Area', Press Release, PR/OSMR/300713 (30 July 2013).

[37]UN, *S/2008/728*, 14.

a peace agreement, the local population and the political process. Used in the right way, deterrence is making the actual use of force needless, which is very much in line with the norms of peacekeeping.

Deterrence was indeed considered in both MONUC and MONUSCO, particularly in the protection of the political process as well as civilians. It was mentioned for the first time in 2004. In August, the Secretary-General suggested that MONUC's military forces should be used to 'deter [...] reprisal attacks by foreign elements against the Congolese civilian population.'[38] In the following resolution, in October the same year, it was decided that MONUC would have the mandate 'to deploy and maintain a presence in the key areas of potential volatility in order to [...] discourage violence, in particular by deterring the use of force to threaten the political process'.[39]

Deterrence is, however, not properly discussed in the mandates, concepts of operations or reports of the Secretary-General until the proposal of the deployment of an EU-led standby force, EUFOR, during the national electoral process in 2006. According to the proposal from the Secretary-General, the '[European Union] force will contribute to stability through a visible advance element and an over-the-horizon force to provide reassurance and deterrence, directly through support for MONUC and indirectly through support for FARDC [the Armed Forces of the Democratic Republic of the Congo] and the national police'.[40] When EUFOR left six months later, the threat of force as a deterrent was much stronger emphasised in resolutions and concepts of operations. In the Security Council's resolution in May 2007, for example, it was decided that MONUC should have the mandate to: 'Deter any attempt at the use of force to threaten the political process from any armed group, foreign or Congolese, particularly in the Eastern part of the Democratic Republic of the Congo'.[41]

Deterrence continues to be important throughout the two missions. In 2013, the Security Council authorised the deployment of a Force Intervention Brigade (FIB), as well as the deployment of a total of around 20 000 forces, providing MONUSCO with substantial military means. In the same resolution, the Council was 'underlining the importance of MONUSCO deterring any threats to the implementation of its mandate'.[42] In subsequent Security Council resolutions and reports from the Secretary-General, it was also acknowledged that MONUSCO 'has successfully deterred attacks on civilians',[43] and that its 'framework brigades' 'must also play a more active

[38]UN, *S/2004/650: Third Special Report of the Secretary-General on the United Nations Organization Mission in the Democratic Republic of the Congo* (16 August 2004), 21.

[39]UN, *S/RES/1565 (2004): Resolution 1565 (2004)* (1 October 2004), 2.

[40]UN, *S/2006/390: Twenty-First Report of the Secretary-General on the United Nations Organization Mission in the Democratic Republic of the Congo* (13 June 2006), 6; UN, *S/RES/1671 (2006): Resolution 1671 (2006)* (25 April 2006).

[41]UN, *S/RES/1756 (2007)*, 4.

[42]UN, *S/RES/2098 (2013): Resolution 2098 (2013)* (28 March 2013), 4.

[43]UN, *S/RES/2147 (2014): Resolution 2147 (2014)* (28 March 2014).

role in protecting civilians by deterring [...] armed groups from inflicting violence on the population.'[44] Hence, deterrence was also explicitly articulated as a strategy in these two missions.

Compellence

Military forces can be used not only for protecting the status quo but also to coerce or force an enemy to actively change its behaviour. The logic of compellence is similar to that of deterrence, but instead of using the threat of force to deter an enemy from taking action, the threat of force is used to persuade an enemy to stop or undo an action it is already undertaking, or to give up something of value. By using threats or limited military force, the coercer increase the costs of the enemy's preferred action and persuade it to behave in a way it would otherwise not do.[45]

Although peacekeeping operations are generally deployed after a ceasefire is established and a peace agreement is signed, the deployment of peacekeepers in violent environments makes it necessary for them to sometimes challenge the status quo. Compellence can therefore also be a useful strategy for peacekeeping operations.[46] First, the threat of force to compel spoilers to *stop their actions* could be valuable in, for example, the protection of civilians. When the belligerents adhere to a ceasefire or a peace agreement, deterrence can be used to discouraging them from using violence in the future. But, when spoilers are targeting the civilian population, a strategy of compellence is needed to stop them from attacking. While defence is focused on the physical protection of the civilian population, i.e., on making it physically impossible for spoilers to succeed with their attacks, compellence is focused on how the threat of force can change their willingness to attack. By threatening to use force against the spoilers when they do not cease to attack civilians, the costs of continued fighting increases for the spoilers.

Second, the threat of force to compel spoilers to *give up something of value* could also be useful in peacekeeping operations, for example, when they do not disarm voluntarily. The disarmament and demobilisation of ex-combatants is often part of a peace agreement, and is usually included as one of the objectives of a peacekeeping operation. The ex-combatants are,

---

[44]UN, *S/2014/957: Report of the Secretary-General on the United Nations Organization Stabilization Mission in the Democratic Republic of the Congo Submitted Pursuant to Paragraph 39 of Security Council Resolution 2147 (2014)* (30 December 2014), 12.

[45]For a detailed discussion on compellence, and coercive diplomacy in interventions, see Bellamy, 'Lessons Unlearned'; Jakobsen, *Western Use of Coercive Diplomacy*; Ohnishi, 'Coercive Diplomacy and Peace Operations'.

[46]This is also true for closely related operations, such as military operations in refugee disasters. According to Posen, 'rescuers' in refugee disasters 'more often find themselves in the active compellence mode than the deterrent mode'. Posen, 'Military Responses to Refugee Disasters', 81.

however, often highly reluctant to voluntarily give up their only means of protection. Here, strategies of brute force (see the discussion below) or compellence could be useful for the peacekeepers. By using the threat of force to compel the spoilers to disarm, non-compliance becomes too costly.

There are a few examples of the use of compellence to disarm the spoilers in the two missions. The disarmament of armed groups was one of the objectives of MONUC already in 2001 but was initially considered in voluntary terms.[47] The military forces were only intended to be used in the defence of the disarmament sites. The use of compellence for the disarmament of armed groups did, however, increase over the years. In 2013, for example, MONUSCO set a deadline for one of the Congolese rebel groups, the March 23 Movement (M23), to disarm. In a press release published on the 30 July 2013, it was stated that:

> MONUSCO will support the FARDC in establishing a security zone in Goma and its northern suburbs [...]. Any individuals in this area who are not members of the national security forces will be given 48 hours as of 4pm (Goma time) on Tuesday 30 July to hand in their weapon to a MONUSCO base and join the DDR/RR process. After 4pm on Thursday 1 August, they will be considered an imminent threat of physical violence to civilians and MONUSCO will take all necessary measures to disarm them, including by the use of force in accordance with its mandate and rules of engagement.[48]

Another example is when the Team of International Envoys, including the UN Special Representative and Head of MONUSCO, threatened to use force against a Rwandan rebel group, the Democratic Forces for the Liberation of Rwanda (FDLR), residing in eastern DR Congo. In 2015, the Team released a press statement concerning the failing of the FDLR to comply with a deadline previously set by the International Conference on the Great Lakes Region (ICGLR) and the Southern African Development Community (SADC) for 'the full and unconditional surrender and demobilization of the Democratic Forces for the Liberation of Rwanda (FDLR).'[49] They stated that:

> By failing to fully comply with the decisions of the ICGLR, SADC and the United Nations Security Council, the FDLR has left the region and the international community with no other option than to pursue the military option against those within the armed group that unwilling to voluntary disarm. [...] The Envoys hereby call upon the DRC Government and MONUSCO, including its Force Intervention Brigade (FIB), to take all necessary measures to disarm the FDLR.[50]

---

[47]UN, *S/RES/1355 (2001): Resolution 1355 (2001)* (15 June 2001).
[48]MONUSCO, 'MONUSCO Deploys to Support Security Zone in Goma'.
[49]MONUSCO, 'International Envoys for the Great Lakes Region Call for Decisive Actions against the FDLR', Press Release (2 January 2015).
[50]MONUSCO, 'International Envoys for the Great Lakes Region'.

There are also examples of the use of limited force to compel spoilers to withdraw from territory. In 2008, for example, the Secretary-General reported that 'MONUC took offensive action against CNDP [the National Congress for the Defence of the People] at the outset of the fighting to compel it to return to positions held on 28 August'.[51] Hence, although compellence is an active strategy, challenging the status quo, it was clearly considered in the mandates and concepts of operations of the two missions.

Offence

Finally, we have strategies of offence or brute force, i.e., strategies with the aim of defeating an enemy. Instead of aiming at persuading an enemy to comply, offence seeks to impose compliance on the enemy. The enemy stops fighting, not because the costs are considered too high compared to the benefits of continued fighting, but because they do not have the means to continue. Offence is furthest from the peacekeeping norm. Nevertheless, when disarming spoilers is one of the objectives, strategies of offence can be highly useful.

First, the most obvious way in which offence can be useful in the disarmament of spoilers is through direct military operations. By launching military operations against the headquarters, bases and fighters, either unilaterally or together with the host state, the capacity of the spoilers can be reduced.

In the DR Congo, one of the first traces of the use of brute force by MONUC was in the mid-2004, when spoilers in eastern DR Congo managed to seize the town of Bukavu in South Kivu.[52] This event forced MONUC to change its strategy in the eastern parts of the country, and in October the same year, MONUC was authorised to forcefully disarm foreign armed groups.[53] Instead of relying on compellence, trusting the spoilers to voluntarily give up their weapons, MONUC was now supposed to 'take a more active and robust role' in disarming the foreign armed groups, including measures such as 'cordon and search operations.'[54] The Security Council also authorised the rapid deployment of two additional infantry battalions and four attack helicopters for these purposes.[55]

Although there are some references to deterrence in the mandates and concepts of operations, the main focus has been on limiting the physical capability of the spoilers. With the deployment of two brigades in the North

---

[51]UN, *S/2008/728*, 11.
[52]UN, *S/2004/650*.
[53]UN, *S/RES/1565 (2004)*.
[54]UN, *S/2004/650*, 21.
[55]UN, *S/RES/1565 (2004)*. This was specifically asked for in a letter from the Secretary-General to the President of the Security Council. UN, *S/2004/715: Letter dated 3 September 2004 from the Secretary-General addressed to the President of the Security Council* (7 September 2004).

and South Kivu in early 2005, for example, it was emphasised that MONUC would 'step up military pressure by conducting operations to disrupt and weaken FDLR formations and thus limit the space within which it can operate',[56] as well as to conduct military operations in FDLR-held areas in North Kivu to 'limit the group's freedom of movement.'[57]

In 2009, the Secretary-General used the word 'neutralization' for the first time, clearly indicating the physical destruction of the spoilers. In one of his reports, the Secretary-General stated that MONUC had assisted FARDC in follow-up operations against FDLR, 'pursuing the neutralization of FDLR'.[58] In 2013, the UN Security Council authorised the deployment of an 'Intervention Brigade' with 'the responsibility of neutralizing armed groups', authorising targeted offensive operations for the first time.[59] The Intervention Brigade, consisting of three infantry battalions, one artillery and one Special Force and Reconnaissance Company, was intended to 'carry out targeted offensive operations' in order 'to prevent the expansion of all armed groups,' and to 'neutralize these groups'.[60]

Second, another way in which brute force can be valuable in the disarmament of spoilers is by *indirectly* limiting their access to resources. Arms embargoes and sanctions against countries that provide support to the belligerents of a conflict are often used in parallel with peacekeeping operations. Peacekeeping forces can be used to enforce these sanctions, thereby decreasing the spoilers' access to resources, focusing on destroying their physical ability to continue fighting.

In DR Congo, arms embargoes and sanctions were indeed used parallel to MONUC and MONUSCO. In UNSCR 1493 in 2003, the Security Council decided that

> all States, including the Democratic Republic of the Congo, shall [...] take the necessary measures to prevent the direct or indirect supply, sale or transfer, from their territories or by their nationals, [...] of arms and any related materiel, and the provision of any assistance, advice or training related to military activities, to all foreign and Congolese armed groups and militias operating in the territory of North and South Kivu and of Ituri [...].[61]

Initially, MONUC was only intended to monitor and observe the compliance with the arms embargo.[62] In 2004, it was, however, stated in the concept of operations that MONUC 'would take a more active and robust role' in the

---

[56]UN, *S/2005/167: Seventeenth Report of the Secretary-General on the United Nations Organization Mission in the Democratic Republic of the Congo* (15 March 2005), 9.
[57]UN, *S/2005/506: Eighteenth Report of the Secretary-General on the United Nations Organization Mission in the Democratic Republic of the Congo* (2 August 2005), 7.
[58]UN, *S/2009/160*, 3.
[59]UN, *S/RES/2098 (2013)*, 6.
[60]UN, *S/RES/2098 (2013)*, 6–7.
[61]UN, *S/RES/1493 (2003): Resolution 1493 (2003)* (28 July 2003), 4.
[62]UN, *S/RES/1291 (2000)*.

disarmament of armed groups, through measures such as 'declaration of weapon-free zones and operations to ensure respect for the arms embargo, with a view to preventing the resupply of the foreign armed groups, from whatever source.'[63] Although brute force is not used against third parties providing support to the spoilers, it is focused on making the spoilers unable to continue fighting.

Problems with contemporary strategy

Although explicit traces of all four strategies can be found in both MONUC and MONUSCO, evidence from the two operations suggests that the potential of military strategy is not fully utilised.

First, there is no coherent, main strategy for how the military forces should be used. Although we see a shift from self-defence towards coercion and offence over time, traces of all fours strategies can be found in both missions. In a Security Council resolution from 2014, for example, the Security Council was '*underlining* the importance of MONUSCO deterring any threats to the implementation of its mandate',[64] simultaneously authorising MONUSCO to take all necessary measures to '[e]nsure, within its area of operations, effective protection of civilians under threat of physical violence',[65] and to 'prevent the expansion of all armed groups, neutralize these groups, and disarm them'.[66]

Furthermore, in one of the mandates from 2008, intentions of deterrence, defence and offence can all be traced within one paragraph. MONUC was authorised to:

> Deter any attempt at the use of force to threaten the Goma and Nairobi processes from any armed group [...] including by using cordon and search tactics and undertaking all necessary operations to prevent attacks on civilians and disrupt the military capability of illegal armed groups.[67]

But, in order to be effective, different strategies require different means (i.e., forces and weapons). According to, for example, Snyder, the capacity for deterrence does not necessarily correspond with the capacity for fighting wars effectively and cheaply, and a certain force posture 'might produce strong deterrent effects [but] not provide a very effective denial and damage-alleviating capability.'[68] This could also be the case for peacekeepers. By optimising peacekeeping forces for defence, for example, they

---

[63]UN, *S/2004/650*, 21. UN Security Council resolution 1533 authorize MONUC to seize or collect arms. UN, *S/RES/1533 (2004): Resolution 1533 (2004)* (12 March 2004).
[64]UN, *S/RES/2147 (2014)*, 5.
[65]UN, *S/RES/2147 (2014)*, 6.
[66]UN, *S/RES/2147 (2014)*, 7.
[67]UN, *S/RES/1856 (2008): Resolution 1856 (2008)* (22 December 2008), 4.
[68]Snyder, *Deterrence and Defense*, 4.

might not be able to deter spoilers from attacking in the first place. Without a coherent strategy, it becomes highly difficult to ensure the right means for achieving the desired ends.

Second, there are indications that the military strategies of MONUC and MONUSCO were developed in reaction to events on the ground, rather than proactively, leaving the initiative to the spoilers. One example is the disarmament of foreign armed groups. Before 2004, disarmament was voluntary, and the contribution of MONUC's armed forces was to provide protection at the disarmament sites, i.e., they were primarily used in a defensive manner.[69] In a report from 2002, it was explicitly stated that 'MONUC will in no way attempt to forcibly disarm combatants'.[70] But, after one of the main towns in eastern DR Congo was seized in June 2004 by a Congolese rebel group, the Secretary-General argued for an increase in the UN military presence in the area, and that the mission 'would take a more active and robust role in disarmament, demobilization, repatriation, resettlement and reintegration'.[71] In the following Security Council resolution, in October the same year, MONUC was authorised 'to support operations to disarm foreign combatants led by the FARDC'.[72] Instead of only focusing on defence, MONUC forces were now also considered for offensive operations against spoilers.

Although military strategies certainly need to be adjusted depending on the developments on the 'battlefield,' one of the main values of strategy is that it is proactive. According to Hew Strachan, 'strategy is not simply reactive: its role is to direct the war'.[73] A strategy allows for the preparation of forces for things to come, and to choose the terms of when, where and how to 'fight'. By allowing the spoilers to set the terms, the real potential of strategy is not realised.

Finally, one of the main features of successful coercion is that the coercer has the will and capacity to use force.[74] If the coercer is not prepared to use force to enforce its demands, coercion is not only useless, but could result in the deterioration of a conflict. If the spoilers realise that the peacekeepers are not going to carry out their threats, the spoilers are likely to escalate. It is, however, only too evident that the will of using force in peacekeeping operations is largely lacking. According to a UN report on the use of force in the protection of civilians in UN peacekeeping operations between 2010 and 2013, forces that were authorised and equipped to use force only responded in around 20% of the incidents where civilians were under

---

[69]UN, *S/2002/1005: Special Report of the Secretary-General on the United Nations Organization Mission in the Democratic Republic of the Congo* (10 September 2002), 7.
[70]UN, *S/2002/1005*, 6.
[71]UN, *S/2004/650*, 21.
[72]UN, *S/RES/1565 (2004)*.
[73]Strachan, 'Strategy and the Limitations of War', 48.
[74]See, for example, Art, 'To What Ends Military Power?' 6.

imminent physical threat. In cases where armed troops were already on site, the force was almost never used.[75] Although the report only included the use of force in cases related to the protection of civilians, it reflects an unwillingness to use force, even in the most critical of circumstances.

Conclusion

In this article, I have outlined the logic of four military strategies for peace-keeping operations – defence, deterrence, compellence and offence – using traditional strategic theory as a point of departure. The various strategies can be utilised for the most common objectives of peacekeeping operations. Defence can be used in the protection of civilians as well as for self-defence, to ensure that other parts of the peacekeeping mission can continue its work elsewhere. Deterrence can also be used to deter spoilers from resort-ing to violence against the mission or the civilian population, but it can also be used to deter violence against the political process in general. Compellence is useful when the peacekeepers are deployed in violent environments, to stop ongoing violence against civilians or in the disarma-ment process of the spoilers. Lastly, brute force, although furthest from the peacekeeping norm, can deprive the spoilers of the means of continued fighting, either directly through operations against headquarters, bases and fighters, or indirectly, by enforcing embargoes and sanctions against those supporting the spoilers.

I have also traced the presence of the four strategies in the mandates, concepts of operations and reports of the Secretary-General, in two subse-quent peacekeeping operations. It is evident that strategies of defence, coercion and offence are all considered in contemporary peacekeeping. It is also evident, that traditional strategic theory is useful for analysing peace-keeping operations. The analysis did, however, uncover several problems with the military strategy of these operations. There was no coherent, main strategy for how the forces should be used, and the military strategies seemed to have been developed in reaction to developments on the ground, rather than being proactive. This makes the potential of strategy unrealised, leaving room for improvement. Despite including strategies of coercion, the troops were also reluctant to use force when needed, decreas-ing the utility of both deterrence and compellence.

These findings are, however, relying on official documents only. It is possible that internal documents would have revealed more coherent and proactive strategies. Nevertheless, by outlining the logic of each of the four

---

[75]UN, *A/68/787: Evaluation of the Implementation and Results of Protection of Civilian Mandates in United Nations Peacekeeping Operations* (7 March 2014), 7–8. This also in line with Trevor Findlay's findings in 2002 regarding the willingness of using force by UN peacekeepers. Findlay, *The Use of Force in UN Peace Operations*, 355.

strategies, it becomes easier to trace contradictions in how the use of military force is considered to contribute to the achievement of the political ends of a specific peacekeeping operation.

Outlining the logic of the four strategies is also important for the practice of peacekeeping. On the one hand, if the international community continues to be unwilling to use force in peacekeeping operations, while still deploying in environments where there is no peace to keep, the logic of the four strategies suggest that defence would be the preferred strategy. Instead of attempting to deter or compel spoilers to refrain from hostile action (which requires the will and capacity to use force), peacekeepers would be more successful with a defensive strategy, making it physically more difficult for spoilers to achieve their ends, for example, by establishing safe zones or safe areas.[76] Since safe zones and safe areas only include the use of force in defence of well-defined areas, this strategy is also most in line with the norms of peacekeeping. But, a stationary defence needs a large number of forces, and the areas between established safe zones or safe areas are often left exposed. Furthermore, the peacekeepers are more likely to become targets, which in turn could decrease the willingness of the international community to contribute with troops to these kinds of operations.

On the other hand, if the international community is becoming more willing to use force in peacekeeping operations, they should instead focus on different forms of coercive strategies. For coercive strategies to be effective, the peacekeepers need to establish their credibility by showing their willingness to use force early on. Well-equipped, flexible and mobile forces supported by helicopters and fighter jets, for example, could signal their willingness. But, these kinds of forces can also be used for offensive operations, which would likely decrease the inclination of the belligerents to welcome an international peacekeeping force, making the deployment of peacekeepers less likely. Nevertheless, if the military forces are used in the right way, coercive strategies could decrease the actual use of force.

In practice, the military strategy of a peacekeeping operation is ultimately restrained by a great number of factors, for example, the need for consensus in the Security Council, the will and ability of troop-contributing countries, and not least by the motivation of the conflicting parties. Still, this does not reduce the importance of having a logically coherent idea about how military force can contribute to keep the peace, i.e., to have a military strategy, in peacekeeping operations.

Disclosure statement

No potential conflict of interest was reported by the author.

---

[76]Safe zones and safe havens are often utilized in humanitarian interventions. See Posen, 'Military Responses to Refugee Disasters'.

ORCID

Kersti Larsdotter ⓘ http://orcid.org/0000-0001-5843-3878

Bibliography

Art, Robert J., 'To What Ends Military Power?' *International Security* 4/4 (1980), 3–35. doi:10.2307/2626666

Bellamy, Alexander J, 'Lessons Unlearned: Why Coercive Diplomacy Failed at Rambouillet', *International Peacekeeping* 7/2 (2000), 95–114. doi:10.1080/13533310008413837

Boutellis, Arthur J., 'From Crisis to Reform: Peacekeeping Strategies for the Protection of Civilians in the Democratic Republic of the Congo', *Stability: International Journal of Security and Development* 2/3 (2013), article 48, 1–11.

Byman, Daniel and Matthew Waxman, *The Dynamics of Coercion: American Foreign Policy and the Limits of Military Might* (Cambridge: Cambridge University Press 2002).

Edstrom, Hakan and Dennis Gyllensporre, *Political Aspirations and Perils of Security: Unpacking the Military Strategy of the United Nations* (Houndmills: Palgrave Macmillan 2013).

Findlay, Trevor, *The Use of Force in UN Peace Operations* (Oxford: Oxford University Press 2002).

Fortna, Virginia, *Page, Does Peacekeeping Work? Shaping Belligerents' Choices after Civil War* (Princeton: Princeton University Press 2008).

Freedman, Lawrence, 'Ukraine and the Art of Limited War', *Survival* 56/6 (2014), 7–38.

Genocide Prevention Task Force, *Preventing Genocide: A Blueprint for U.S. Policymakers* (Washington, DC: United States Holocaust Memorial Museum 2008).

Gray, Colin S., *Maintaining Effective Deterrence* (Carlisle, PA: SSI, US Army War College 2003).

Hillen, John, *Blue Helmets: The Strategy of UN Military Operations* (London: Brassey's 1998).

Hunt, Charles T., 'All Necessary Means to What Ends? the Unintended Consequences of the "Robust Turn" in UN Peace Operations', *International Peacekeeping* 24/1 (2017), 108–31. doi:10.1080/13533312.2016.1214074

Jakobsen, Peter Viggo, *Western Use of Coercive Diplomacy: A Challenge for Theory and Practice* (Houndmills, Basingstoke, Hampshire: Palgrave Macmillan 1998).

Karlsrud, John, 'The UN at War: Examining the Consequences of Peace-Enforcement Mandates for the UN Peacekeeping Operations in the CAR, the DRC and Mali', *Third World Quarterly* 36/1 (2015), 40–54. doi:10.1080/01436597.2015.976016

Kjeksrud, Stian, "The Utility of Force for Protecting Civilians", in Haidi Willmot, Ralph Mamiya, Scott Scheeran, and Marc Weller (eds.), *Protection of Civilians* (Oxford: Oxford University Press 2016), 329–49.

Larsdotter, Kersti, *Military Interventions in Internal Wars: The Study of Peace or the Study of War?* (Gothenburg: School of Global Studies, University of Gothenburg 2011).

Lonsdale, David J., 'Strategy: The Challenge of Complexity', *Defence Studies* 7/1 (2007), 42–64. doi:10.1080/14702430601135578

Meiser, Jeffrey W., 'Ends + Ways + Means = (Bad) Strategy', *Parameters* 46/4 (2016–17), 81–91.

MONUSCO, 'MONUSCO Deploys to Support Security Zone in Goma – Sake Area', Press Release, PR/OSMR/300713 (30 July 2013). Available at https://monusco. unmissions.org/en/monusco-deploys-support-security-zone-goma-sake-area

MONUSCO, 'International Envoys for the Great Lakes Region Call for Decisive Actions against the FDLR', Press Release (2 January 2015). Available at https://monusco. unmissions.org/node/100043758

Ohnishi, Ken, 'Coercive Diplomacy and Peace Operations: Intervention in East Timor', *NIDS Journal of Defense and Security* 13 (2012), 53–77.

Pape, Robert A., 'Coercion and Military Strategy: Why Denial Works and Punishment Doesn't', *Journal of Strategic Studies* 15/4 (1992), 423–75. doi:10.1080/01402399208437495

Posen, Barry R., *The Sources of Military Doctrine: France, Britain, and Germany between the World Wars* (Ithaca, NY: Cornell University Press 1986).

Posen, Barry R., 'Military Responses to Refugee Disasters', *International Security* 21/1 (1996), 72–111. doi:10.1162/isec.21.1.72

Rosen, Stephen Peter, 'Vietnam and the American Theory of Limited War'', *International Security* 7/2 (1982), 83–113. doi:10.2307/2538434

Schelling, Thomas C., *Arms and Influence* (New Haven, CT: Yale University Press 1966).

Seybolt, Taylor B, *Humanitarian Military Intervention: The Conditions for Success and Failure* (Oxford: Oxford University Press 2008).

Snyder, Glenn H., *Deterrence and Defense: Toward a Theory of National Security* (Princeton, NJ: Princeton University Press 1961).

Strachan, Hew, 'Strategy and the Limitations of War', *Survival* 50/1 (2008), 31–54. doi:10.1080/00396330801899470

Strachan, Hew, 'Strategy or Alibi? Obama, McChrystal and the Operational Level of War', *Survival* 52/5 (2010), 157–82. doi:10.1080/00396338.2010.522104

Tardy, Thierry, 'A Critique of Robust Peacekeeping in Contemporary Peace Operations', *International Peacekeeping* 18/2 (2011), 152–67. doi:10.1080/13533312.2011.546089

UN, *A/68/787*, '*Evaluation of the Implementation and Results of Protection of Civilian Mandates in United Nations Peacekeeping Operations*', (7 March 2014).

UN, *S/1999/790*, '*Report of the Secretary-General on the United Nations Preliminary Deployment in the Democratic Republic of the Congo*', (15 July 1999).

UN, *S/2000/30*, '*Report of the Secretary-General on the United Nations Organization Mission in the Democratic Republic of the Congo*', (17 January 2000).

UN, *S/2001/128*, '*Sixth Report of the Secretary-General on the United Nations Organization Mission in the Democratic Republic of the Congo*', (12 February 2001).

UN, *S/2001/572*, '*Eighth Report of the Secretary-General on the United Nations Organization Mission in the Democratic Republic of the Congo*', (8 June 2001).

UN, *S/2002/1005*, '*Special Report of the Secretary-General on the United Nations Organization Mission in the Democratic Republic of the Congo*', (10 September 2002). doi:10.1044/1059-0889(2002/er01)

UN, *S/2002/621*, '*Eleventh Report of the Secretary-General on the United Nations Organization Mission in the Democratic Republic of the Congo*', (5 June 2002). doi:10.1044/1059-0889(2002/er01)

UN, *S/2004/1034*, 'Sixteenth Report of the Secretary-General on the United Nations Organization Mission in the Democratic Republic of the Congo', (31 December 2004).

UN, *S/2004/650*, 'Third Special Report of the Secretary-General on the United Nations Organization Mission in the Democratic Republic of the Congo', (16 August 2004).

UN, *S/2004/715*, 'Letter dated 3 September 2004 from the Secretary-General addressed to the President of the Security Council', (7 September 2004).

UN, *S/2005/167*, 'Seventeenth Report of the Secretary-General on the United Nations Organization Mission in the Democratic Republic of the Congo', (15 March 2005).

UN, *S/2005/506*, 'Eighteenth Report of the Secretary-General on the United Nations Organization Mission in the Democratic Republic of the Congo', (2 August 2005).

UN, *S/2006/390*, 'Twenty-First Report of the Secretary-General on the United Nations Organization Mission in the Democratic Republic of the Congo', (13 June 2006).

UN, *S/2008/728*, 'Fourth Special Report of the Secretary-General on the United Nations Organization Mission in the Democratic Republic of the Congo', (21 November 2008).

UN, *S/2009/160*, 'Twenty-Seventh Report of the Secretary-General on the United Nations Organization Mission in the Democratic Republic of the Congo', (27 March 2009).

UN, *S/2014/957*, 'Report of the Secretary-General on the United Nations Organization Stabilization Mission in the Democratic Republic of the Congo Submitted Pursuant to Paragraph 39 of Security Council Resolution 2147 (2014)', (30 December 2014).

UN, *S/RES/1291 (2000)*, 'Resolution 1291 (2000)', (24 February 2000).

UN, *S/RES/1355 (2001)*, 'Resolution 1355 (2001)', (15 June 2001).

UN, *S/RES/1493 (2003)*, 'Resolution 1493 (2003)', (28 July 2003).

UN, *S/RES/1533 (2004)*, 'Resolution 1533 (2004)', (12 March 2004).

UN, S/RES/1565 (2004), 'Resolution 1565 (2004)', (1 October 2004).

UN, *S/RES/1671 (2006)*, 'Resolution 1671 (2006)', (25 April 2006).

UN, *S/RES/1756 (2007)*, 'Resolution 1756 (2007)', (15 May 2007).

UN, *S/RES/1856 (2008)*, 'Resolution 1856 (2008)', (22 December 2008).

UN, *S/RES/2098 (2013)*, 'Resolution 2098 (2013)', (28 March 2013).

UN, *S/RES/2147 (2014)*, 'Resolution 2147 (2014)', (28 March 2014).

UN, S/RES/2277 (2016), 'Resolution 2277 (2016)', (30 March 2016).

Williams, Paul D., *Enhancing Civilian Protection in Peace Operations: Insights from Africa* (Washington, DC: Africa Center for Strategic Studies 2010).

Fancy bears and digital trolls: Cyber strategy with a Russian twist

Benjamin Jensen, Brandon Valeriano and Ryan Maness

ABSTRACT
How states employ coercion to achieve a position of advantage relative to their rivals is changing. Cyber operations have become a modern manifestation of political warfare. This paper provides a portrait of how a leading cyber actor, Russia, uses the digital domain to disrupt, spy, and degrade. The case illustrates the changing character of power and coercion in the twenty-first century. As a contribution to this special issue on twenty-first century military strategy, the findings suggest new forms of competition short of war.

In October 2017, major social media firms including Facebook revealed that targeted Russian propaganda during the 2016 US presidential election may have reached as many as 126 million users.[1] These ads, tailored to "unleash the protest potential of the population"[2] formed a new front in a long-term competitive strategy designed to undermine US institutions and resolve. This information warfare campaign, combining propaganda with cyber intrusions, reflects a twenty-first century form of political warfare.[3]

The power to hurt has become the power to hurt online.[4] Just as the nuclear age heralded important changes to conceptualizing the use of force to achieve political objectives, the connectivity of the twenty-first century alters how rival states seek a position of relative advantage and coerce their

---

[1] Mike Isaac and Daisuke Wakabayashi, 'Russian Influence Reached 126 Million Through Facebook Alone', *New York Times*, 30 Oct. 2017.

[2] Gerasimov, 'The Value of Science in Prediction', *Military-Industrial Kurier* (27 Feb. 2013).

[3] George F. Kennan on Organizing Political Warfare, 'History and Public Policy Program Digital Archive, Obtained and contributed to CWIHP by A. Ross Johnson', Cited in his book *Radio Free Europe and Radio Liberty*, Ch1 n4 – NARA release courtesy of Douglas Selvage. Redacted final draft of a memorandum dated 4 May 1948, and published with additional redactions as document 269, FRUS, Emergence of the Intelligence Establishment, 30 Apr. 1948.

[4] Erik Gartzke and Jon R. Lindsay, 'Coercion through the Cyberspace: The Stability-Instability Paradox Revisited', in Kelly Greenhill and Peter Krause (eds.), *The Power to Hurt in the Modern World* (Oxford: Oxford University Press 2017).

adversaries. Coercion, the exploitation of potential force short of war,[5] is reborn as disruptive website defacements and denial of service attacks, massive espionage campaigns, deception, and covert psychological warfare designed to shape decisions in rival states.[6]

This paper investigates how Russia employs cyber ways and means to achieve strategic ends. As a contribution to this special issue on twenty-first century military strategy, the paper explores how rival states employ cyberspace to achieve a relative position of advantage and shape their opponents' decision architecture. The Kremlin is not alone in employing cyber coercive instruments as part of long-term competition between rivals. Great powers use any means at their disposal to advance their interests. The US and Israel show a penchant for combining cyber sabotage alongside the threat of military force and other coercive diplomatic instruments.[7] China illustrates how espionage and deception alter the long-term balance of information in incidents such as the Office of Personnel Management intrusion and large-scale intellectual property theft.[8] Even regional actors such as North Korea show how cyberspace can be used to coerce firms, illicitly access hard currency, and spy on rivals.[9] Here, we offer a portrait of one of the leading cyber actors, Russia, and how it employs a mix of coercion and espionage to advance their its interests online.

Cyber strategy has come of age. Strategy is a dialectic of opposing wills that revolves around a set of ideas about how to employ instruments of power to advance a defined objective.[10] Strategy therefore is the "art of creating power."[11] For Robert Osgood, this art of power "must be understood as nothing less than all plans for utilizing the capacity for armed coercion – in conjunction with the economic, diplomatic, and psychological instruments of power – to support foreign policy most effectively by overt, covert, and tacit means."[12] Strategy therefore is a guide to long-term competition and this struggle, by definition, involves interdependent decisions and expectations about rival behavior.[13]

[5]Thomas Schelling, *Strategy of Conflict* (Cambridge: Harvard University Press 1960), 9.
[6]Brandon Valeriano, Benjamin Jensen, and Ryan Maness, *Cyber Strategy: The Changing Character of Cyber Power and Coercion* (New York: Oxford University Press 2018).
[7]On Stuxnet, see Jon R. Lindsay, 'Stuxnet and the Limits of Cyber Warfare', *Security Studies* 22/3 (2013), 365–404; Rebecca Slayton, 'What is the Cyber Offense-Defense Balance? Conceptions, Causes, and Assessments', *International Security* 41/3 (2016), 72–109.
[8]Jon R. Lindsay, 'The Impact of China on Cybersecurity: Fiction and Friction', *International Security* 39/3 (2014), 7–47.
[9]On the Sony Pictures Hack, see Travis Sharp, 'Theorizing Cyber Coercion: The 2014 North Korea Operation against Sony', *Journal of Strategic Studies* 40/7 (2017), 898–926. For an extensive overview of a known APT linked to North Korea, see Kaspersky Labs, *Lazarus Under the Hood* (25 Nov. 2017).
[10]The idea of strategy as a dialectic come from Andre Beaufre, *An Introduction of Strategy* (London: Faber and Faber 1965 R.H. Barry translation), 22.
[11]Lawrence Freedman, 'Strategic Studies and the Problem of Power', in Thomas Mahnken and Joseph A. Maiolo (eds.), *Strategic Studies: A Reader* (New York: Routledge 2008), 31.
[12]Robert Osgood, *The Entangling Alliance* (Chicago: University of Chicago Press 1962).
[13]Schelling, *Strategy of Conflict*, 3.

In Clauswitzean terms, strategy, as the art of creating power, has an enduring nature and a changing character. The enduring feature is the competitive struggle against an adversary using all available means directed against an, ideally, clear political objective. The changing character resides in the prevailing theories of victory and available resources actors can use to achieve a position of advantage.[14] From political institutions to new technology and changing social norms, interconnected factors define strategic practice.[15] As more and more of our daily lives occur in a digital space, the logic of strategy shifts to a new domain.

To investigate how Russia employs cyber strategy in pursuit of political objectives, this paper proceeds as follows. First, we situate cyber strategy within the broader theoretical literature on coercion and covert signaling. Second, we review documented cyber operations pursued by Moscow. These operations show that most cyber intrusions focus on disruption and harassment alongside espionage required to gain access and collect intelligence. Although Russia has employed cyber instruments alongside major combat operations in Georgia and Ukraine, the preponderance of Russia activity is disruptive rather than degrading. The paper concludes by noting that although Russian cyber information operations are concerning, their efficacy is questionable and these operations represent the actions and declining power.

The character of cyber strategy

Most cyber intrusions between great powers and rival states do not involve wartime exchanges.[16] Rather, they take place in what defense pundits are increasing calling a gray zone short of war. In this respect, the use of cyber operations, defined as the use of malicious code to alter or destroy information or physical networks, is similar to covert action. They reflect concealed means to achieve a political end.

Rival states seek to compel one another and manage escalation risks through a variety of instruments in the shadows. From Sun Tzu and Kautilya to Thomas Schelling and Alexander George, covert action and coercion are major themes in strategic and military theory. As a form of hostile covert action, cyber operations can represent ambiguous signals designed to probe adversary intentions and manage escalation risk. The reality is that cyber operations do not produce concessions in isolation. Instead, they often seek to distract an opponent or amplify a propaganda theme. Furthermore, most cyber operations

---

[14]On theories of victory, see Benjamin Jensen, *Forging the Sword: Doctrinal Change in the U.S. Army* (Palo Alto: Stanford University Press 2016).
[15]Beatrice Heuser, *The Evolution of Strategy: Thinking War from Antiquity to the Present* (New York: Cambridge University Press 2010), 19–24.
[16]For empirical data on rival state use of cyber, see Valeriano, Jensen and Maness, *Cyber Strategy*.

involve espionage and shaping activities required to understand adversary networks and gain access. The traditional understanding of coercion, as articulated by Thomas Schelling, finds resonance in cyber operations, but only with weak effects.[17] Compellence is rare, deterrence uncertain in a realm of covert action.

Cyber operations, in addition to their latent intelligence value, can act as additive measures that amplify existing strategic signals. Their coercive effect, at least to date, is more latent than manifest due to issues associated with attribution and credibility issues. In a crisis, states can also face problems clearly communicating their intent through signals. This challenge is amplified in cyberspace, where "the linkages between intent, effect, and perception are loose."[18] This dynamic creates a condition in which signals "can be as or more ambiguous when they take place or refer to events in cyberspace than they are when limited to the physical world."[19] There are also unique signaling challenges associated with cyber coercion that limit its power to hurt. According to Borghard and Lonergan, "signaling in cyber space is the problematic of all domains (land, sea, air, space and cyber) because the signal may go unrealized. In other words, in cyberspace only the initiator may perceive the engagement."[20] Similarly, for Gartzke and Lindsay,

> The biggest obstacle to cyber coercion is the difficulty of credibly signaling about potential harm that depends on secrecy to be harmful... Sacrifice of anonymity on which offensive deception depends exposes the cyber attacker to retaliation. Coercive cyber threats thus tend to be more generalized, which undercuts their effectiveness in targeted or crisis situations.[21]

As concealed means, cyber operations sacrifice signal strength for network access but gain benefits from anonymity.

If strategy is a concept concerning how to influence rivals in pursuit of political objectives, then what forms of interaction help states create the power to do so in the digital domain? We propose three distinct strategic logics in cyberspace: disruption, espionage, and degradation.[22] These logics build on earlier work on coercive and coercion as well as recent explorations of cyber coercion, but with an important caveat.[23] Cyber strategy need not seek a direct concession and tends to occur predominantly in the covert, as opposed to overt, space. Rival states use indirect cyber instruments to shape long-term competition more than they seek immediate concessions. Russia

---

[17]Schelling, *Strategy of Conflict*.
[18]Martin Libicki, *Crisis and Escalation in Cyberspace* (Santa Monica: Rand Corporation 2012), xvi.
[19]Libicki, *Crisis and Escalation in Cyberspace*, xv.
[20]Erica Borghard and Shawn Lonergan, 'The Logic of Coercion in Cyberspace', *Security Studies* 26/3 (2017), 452–481.
[21]Gartzke and Lindsay, 'Coercion through the Cyberspace', 26.
[22]Valeriano, Jensen, Maness, *Cyber Strategy*.
[23]On coercive diplomacy, coercion, and cyber coercion as they relate to one another, see Valeriano, Jensen and Maness, *Cyber Strategy*.

utilizes destabilizing hacks to harass targets toward bending to Moscow's will. As a coercive tool available to states, cyber operations therefore represent a weak form of coercive diplomacy. Digital intrusions are meant to be used with other sticks and carrots to shape an adversary's decision-making.

Cyber disruptions are a low-cost, low-payoff form of cyber strategy designed to shape the larger bargaining context. These cheap signals likely do not achieve sufficient leverage to compel a target.[24] Rather, they seek to probe an adversary: testing their resolve, signaling escalation risk, and supporting larger propaganda efforts. Website defacements and distributed denial of service (DDoS) incidents are a form of tacit bargaining. According to George Downs and David Rocke, "tacit bargaining takes place whenever a state attempts to influence the policy choices of another state through behavior, rather than by relying on formal or informal diplomatic exchanges [alone]."[25] Low-cost cyber disruptions pressure a rival, through either signaling the risk of crisis escalation or, in combination with propaganda efforts, undermining public confidence in existing policy preferences. Website defacements often echo particular narratives designed to limit policy options for a rival, portraying the opposition as extreme versions of evil, for example how website defacements characterize the Ukrainian government as fascists or Nazis.

As a strategy, cyber espionage concerns altering the balance of information to achieve a position of advantage. Activities can range from simple network penetration to retrieve information to manipulating data to corrupt a rival's confidence in their own systems. These actions are not coercive in the traditional sense. Rather, they concern long-term competition and how rival states seek to find ways of exploiting information asymmetries. Espionage represents efforts to steal critical information or manipulate information asymmetries in a manner that produces bargaining benefits between rival states engaged in long-term competition.

Cyber degradation – coercive operations designed to sabotage the enemy target's networks, operations or systems – is more likely to have a compellent effect than disruptions or espionage. Yet, this effect is rare because many times the target is hardened or too complex to be knocked out for extended periods as a result of malicious cyber actions. This form of cyber strategy resembles denial coercion used in airpower and tends to exhibit sunk costs due to its complexity and tailored design (optimized for a specific system and to achieve destructive effects).[26] This high-cost, high-payoff dynamic makes

---

[24]James D. Fearon, 'Rationalist Expectations for War', *International Organization* 49/3 (1995), 379–414; James D. Fearon 'Signaling Foreign Policy Interests Tying Hands versus Sinking Costs', *Journal of Conflict Resolution* 41/1 (1997), 68–90.

[25]George Downs and David Rocke Downs,*Tacit Bargaining: Arms Races, Arms Control* (Ann Arbor: University of Michigan Press 1990), 3.

[26]Robert Pape, *Bombing to Win: Air Power and Coercion in War* (Ithaca: Cornell University Press 1996).

degradation a costlier signal and thus more likely to achieve effects, but the results are complicated when examined carefully.[27] The Stuxnet operation launched against Iran, similar unsuccessful "left-of-launch" actions directed toward North Korea to prevent them from advancing its missile program, or even actions against Russia in response to the election hacks of 2016 are all examples of cyber degradation.

Cyber strategy therefore can be thought of as a modern variant of coercive diplomacy,

> a political-diplomatic strategy that aims to influence a [rival's] will or incentive structure. It is a strategy that combines threats of force, and if necessary, the limited and selective use of force in discrete and controlled increments, in a bargaining strategy ... the aim is to induce an adversary to comply with one's demands, or to negotiate the most favorable compromise possible, while simultaneously managing the crisis to prevent unwanted military escalation.[28]

Unlike traditional perspectives on coercion, coercive diplomacy can involve positive inducements and is not singular focused on producing concessions (i.e., compellence) or stopping an action before it occurs (i.e., deterrence). Cyber strategies, like coercive diplomacy, are much broader than traditional perspectives on coercion. While rival states can and, as we show, do use cyber operations to compel, they more often than not use the digital domain to signal, steal, and engage in covert propaganda as a means of shaping long-term competition. These shaping operations form the foundation of Russian cyber strategy.

Cyber espionage: access, manipulation, and control

The Russian approach to cyber espionage involves not just stealing critical information but also leveraging it for propaganda value and signaling resolve as a means of changing the trajectory of future crises. By itself, cyber espionage does not achieve concessions. Rather, it is an additive means of accessing networks for future coercion and stealing sensitive information. Used in conjunction with broader propaganda campaigns, espionage can gain access to influence public opinion. These shaping actions do not achieve independent concessions but set the conditions for future crisis bargaining.

Espionage often runs parallel to broader manipulation efforts or set the conditions to follow on actions. One of Russia's cyber espionage toolkits, known as Snake/Uroburo/Tula, first appeared in 2005, targeting systems in the US,

---

[27]On how covert action can signal resolve through sinking costs, see Austin Carson and Keren Yarhi-Milo, 'Covert Communication: The Intelligibility and Credibility of Signaling in Secret', *Security Studies* 26/1 (2017), 124–156.

[28]Jack Levy, 'Deterrence and Coercive Diplomacy: The Contributions of Alexander George', *Political Psychology* 29/4, 539.

United Kingdom, and other Western European countries.[29] Beginning in mid-2013, Operation Armageddon, a cyber espionage campaign that relied predominantly on spearfishing, targeted Ukrainian security services. Spearfishing involves targeted e-mails and communications designed to lure a person into making their machine vulnerable to exploitation. The timing of the attack coincided with the final negotiations between Ukraine and the European Union Association Agreement.[30] Another Russia-linked cyber espionage campaign from a group known as Sandworm surfaced in 2009 based on zero-day exploits affecting Windows operating systems.[31] A zero-day exploit takes advantage of security vulnerabilities inherent in core software or hardware that has yet to be patched. In October 2014, Sandworm used BlackEngery 3 for multiple intrusions in Ukraine focusing on power companies and media outlets.[32] In 2016, Ukrainian power companies along with the finance and defense ministry reported temporary disruptions linked by iSight Partners and attributed them to Sandworm.[33] Espionage efforts such as these reflect how cyber espionage has dual use as a signaling mechanism. The intrusion both accesses critical information that may aid in future cyber coercive incidents as a crisis escalates and signals, ambiguously enough to limit retaliation, to promote Russian interests in external actors.

Cyber espionage is also a means of manipulation and undermining the institutions of opponents. A Russian group known as APT28, or Fancy Bear, similarly used malware to target groups of interest to the Russian state, including security ministries and journalists across the Caucasus region, the Polish and Hungarian governments, NATO, and the Organization for Security and Cooperation in Europe.[34] Unlike traditional Russia cyber-criminal groups, "APT 28 does not exfiltrate financial information from targets and it does not sell the information that it gathers for profit."[35] First discovered in 2011, Energetic Bear is a team that uses a common malware suite to infiltrated networks in the commercial space with economic or defense interests. Initially, the malware appeared on the networks of firms associated with the aviation industry and major defense contractors in the US and Canada. In 2013, the malware appeared on major energy firms such as Exxon-Mobil and British Petroleum.[36] Of note, Energetic Bear is "uniquely

---

[29]BAE Systems, *The Snake Campaign* (Feb. 2014); David Sanger and Steven Erlanger, 'Suspicion Falls on Russia as Snake Cyberattacks Target Ukraine's Government', *The New York Times*, 9 Mar. 2014.

[30]Brian Prince, '"Operation Armageddon" Cyber Espionage Campaign Aimed at Ukraine', *Security Week*, 28 Apr. 2015.

[31]Kim Zetter, 'Russian Sandworm Hack Has Been Spying on Foreign Governments for Years', *Wired*, (14 Oct. 2012.

[32]John Hultquist, *Sandworm team and the Ukrainian Power Authority Attacks* (FireEye 7 Jan. 2016).

[33]Pavel Polityuk, *Ukraine Investigates Suspected Cyber-attack on Kiev Power Grid* (Reuters 20 Dec. 2016).

[34]Threat Intelligence, *APT28: A Window into Russia's Cyber Espionage Operations* (FireEye 27 Oct. 2014).

[35]James Scott and Drew Spaniel, *Know Your Enemies 2.0* (Institute for Critical Infrastructure Technology 2016).

[36]MSS Global Threat Response, *Emerging Threat: Dragonfly/Energetic Bear – APT Group* (Symantec 30 Jun. 2014).

positioned to assist in a combination of digital and physical warfare for military or political purposes."[37] According to F-Secure, since 2008, MiniDuke, along with the CosmicDuke APT group, has acted as state-sponsored espionage organizations.[38] The Duke series first appeared after an 5 April 2008 speech by US President Obama advocating for a missile defense shield in Poland. In 2013, the group was linked to a spear phishing campaign targeting the Ukrainian ministry of foreign affairs.[39]

The style and content of Russian enabled cyber collectives reflect two logics of cyber espionage and its coercive potential. First, accessing target networks sets the conditions for follow-up operations. In military parlance, you prepare the environment for future action. Not only do you access critical networks and steal information altering the balance of information in a crisis, but even if the intrusion is revealed, the target is left wondering what else was stolen and what other networks are compromised.

Second, they demonstrate the utility of cyber espionage as a low-cost means of manipulating public opinion. In this respect, the actions are classic political warfare. Cyber is both a tool to manage crises indirectly and subvert public opinion and political will. In December 2014, "a well-known military correspondent for a large US newspaper was hit via his personal email address in December 2014, probably leaking his credentials. Later that month, Operation Pawn Storm attacked around 55 employees of the same newspaper on their corporate accounts."[40] Linked to APT 28, Pawn Storm infiltrated and disrupted TV5 Monde in France. In October 2015, Pawn Storm set up a fake VPN and fake Outlook Web Access server to conduct spear phishing attacks against the Dutch Safety Board investigating the Malaysian Airlines Flight 17 commercial airline flight that was shot down by a Russian Buk surface-to-air missile over Ukraine.

The manipulation of the 2016 US election

The 2016 US election hack demonstrates how Russia leverages cyber espionage as part of broader active measures campaign.[41] These campaigns do not lay the groundwork for future strikes as much as they focus on altering the perceptions of targeted domestic populations. As such, this event deserves extensive coverage as crucial case of Russian cyber activities. The

---

[37]Scott and Spaniel, *Know Your Enemies 2.0*, 29.
[38]Sarah Peters, 'MiniDuke, CosmicDuke APT Group Likely Sponsored by Russia', in *Dark Reading* (17 Aug. 2015).
[39]Sean Gallagher, 'Seven Years of Malware Linked to Russian State-Backed Cyber Espionage', *Arstechnica*, 17 Sept. 2015.
[40]Feike Hacquebord, *Operation Pawn Storm Ramps Up its Activities, Targets NATO, White House*, (Trend Micro 16 Aug. 2015).
[41]Adam Hulcoop, John Scott-Railton, Peter Tanchak, Matt Brooks, and Ron Deibert, *Tainted Leaks: Disinformation and Phishing With a Russian Nexus* (Citizen Labs May 2017).

goal of the 2016 election hack was to achieve a broader psychological impact on American society while demonstrating to a Russian audience the corrupt nature of democratic institutions. The event serves as a demonstration of the utility of espionage as tool of manipulation. Often termed reflexive control or active measures, this form of espionage tries to influence the adversary through information control and manipulation aided by propaganda, operations all conducted short of war.

The Russian political warfare operation that culminated in the election hack started in 2015, before Donald Trump entered the presidential race. During the summer of 2015, Russia started the process by sending out thousands of phishing emails trying to get their targets to click on malicious links. Thomas Rid during Senate Testimony noted that about 2.4 percent of the attacks were successful in producing information.[42]

The breaches by Russia were not discovered until June 2016 with the *New York Times* noting that two different groups of Russian hackers (Cozy and Fancy Bear) penetrated the Democratic National Committee's (DNC) computer systems.[43] The goal was to monitor the DNC's communications while also exfiltrating their files including opposition research on Donald Trump. This information, in the tradition of KGB, could be used to either augment larger influence operations or as future blackmail material to gain leverage.

Perhaps the most devastating information grabs were the emails taken from Hillary Clinton's staff, further exacerbating a long-standing issue surrounding the question of stored emails on her personal home server. Campaign Chairman John Podesta's emails were stolen due to a typo on advice from an IT consultant to not answer a phishing email (illegitimate was corrected to legitimate).

The Dukes, or Cozy Bear as the Information Security (InfoSec) community calls them, group has been caught before operating in unclassified White House systems, the State Department, and various other US organizations. By the summer of 2015, they started to penetrate both DNC and Republican National Committee files.[44]

In March 2016, Fancy Bear or APT28 piled on. This intrusion thus demonstrates the uncoordinated nature of Russian cyber operations with duplicate processes occurring. Both actors were attacking the same targets, seemingly under the same mandate without overall coordination under similar instructions. Podesta's emails where released the same day as damning audio tape

---

[42]Thomas Rid, 'Disinformation: A Primer in Russian Active Measures and Influence Campaigns', Hearings before the Select Committee on Intelligence, United States Senate, One Hundred Fifteenth Congress, 30 Mar. 2017.

[43]David Sanger, 'D.N.C. Says Russian Hackers Penetrated Its Files, Including Dossier on Donald Trump', *New York Times*, 14 Jun. 2016.

[44]Adam Greenberg, 'Russia Hacked "Older" Republican Emails, FBI Director Says', *Wired*, 10 Jan. 2017.

of Trump remaking on his ability to grab women hit the news cycle (the Billy Bush Bus tape incident).

Coordinated drops of information went on until late in the election cycle, demonstrating a high amount of collaboration between the brokers of information and the Trump campaign. Russian activities seemed to continue until a meeting between Obama and Putin at the G20 on 5 September where Obama warned against further attempts to influence the election. The information dumps stopped but Russian hackers continued to probe state level election voting systems looking for weaknesses.

In July 2016, Clinton's campaign suggested that Russia might be trying to sway the election.[45] Yet, it was not until 7 October that the Obama Administration formally accused Russia of interfering with the election.[46] A bipartisan statement was drafted but the Senate Republicans would not sign on the general statement supporting the noninterference in elections, suggesting it would tip the scales for the Democrats in the election.[47]

The issue was more complicated than simple Russian inference though; the main candidate encouraged the intrusions and information dumps. That Trump was "embracing an unlikely ally" in Wikileaks was certainly a troubling development.[48] Julian Assange, the leader of the organization, blames Hilary Clinton for his predicament and has openly sided with the Russian government, refusing to publish email troves of Russian documents. *Think Progress* counts 164 mentions of Wikileaks by Trump during campaign events.[49]

The coordination through information dumps, botnets supporting the releases, and the mentions by Trump himself demonstrate the power and collaboration needed to make political warfare insidious. As Clint Watts noted in Senate testimony, "part of the reason active measures have worked in the US election is because the Commander-in-Chief has used Russian active measures at times against his opponents."[50] Watts went on to note the many times fake information was released, passed, and amplified by bot networks and then parroted by the Trump campaign itself.

In September 2017, Facebook announced that the company had sold at least $150,000 in ads to Russian operatives after being called into private

---

[45]Eric Lichtblau, 'Computer Systems Used by Clinton Campaign Are Said to Be Hacked, Apparently by Russia', *New York Times*, 20 Jul. 2016.

[46]David Sanger and Charles Savage, 'U.S. Says Russia Directed Hacks to Influence Elections', *New York Times*, 7 Oct. 2016.

[47]Kaveh Waddell, 'Why Didn't Obama Reveal Intel About Russia's Influence on the Election?', *The Atlantic*, 11 Dec. 2016.

[48]Patrick Healy, David Sanger, and Maggie Haberman, 'Donald Trump Finds Improbable Ally in WikiLeaks', *New York Times*, 12 Oct. 2016.

[49]Judd Legum, 'Trump Mentioned WikiLeaks 164 Times in the Last Month of Election, Now Claims it Didn't Impact one Voter', *Think Progress*, 8 Jan. 2017.

[50]Aaron Rupar, 'Former FBI agent Details How Trump and Russia Team Up to Weaponize Fake News', *Think Progress*, 30 Mar. 2017.

questioning by the House of Representatives. "The Agency," a well-known Russian Troll farm, was linked to the ad buys. These ads seemed to seek to influence divisive internal conflict but amplifying issues such as Black Lives Matter. The ads ran from June 2015 to May 2017. As it stands, the *Daily Beast* estimates that, at a minimum, 23 million people saw the ads with a high-end estimate of 70 million.[51] The figure is based on an average of $6 in ad buys results in 1000 views with estimates increasing through targeting and voluntary sharing of the information.

Speculation that Russia was behind the attacks and information releases has been consistent since the issue was first reported in June 2016. A plethora of sources have indicated that the operation was sophisticated and bore the hallmarks of a Russian influence operation, starting with Cloudstrike, a prominent cyber security firm that the DNC turned to. News organizations including the *New York Times*, *Washington Post*, and later, *Politico* all released investigative reports on the operation. Thomas Rid noted in *Esquire* that researchers connected the command server for the malware targeting the DNC to a prior attack on German Parliament in 2015.[52]

In January 2017, the US Intelligence Community as a collective offered their assessment that Russian operations sought to "undermine public faith in the US democratic process, denigrate Secretary [Hillary] Clinton, and harm her electability and potential presidency."[53] The report identified the motive as Putin blaming Clinton personally for the release of the Panama Papers (a series of information dumps locating illicit banking methods) and protests in Russia in 2011 and 2012.

Cyber coercion is difficult, costly, and time-consuming. The Russian operation against the election started well before 2016 and continued past the actual vote. While there is no clear impact that can be detailed, the operation likely reinforced negative opinions of Hillary Clinton already held by Republican voters and supporters of Bernie Sanders. There is no evidence that the operation changed minds, no poll has ever been released that showed support for Trump was generated through Wikileaks. The question is if the hacks motivated individuals to reject Clinton and turn out to vote for Trump.

It appears likely that dozens of strategic mistakes lead to Clinton's loss, including not giving sufficient attention to the Rust belt, the inability to counter fears over immigration, James Comey note that Clinton's emails were under review again, and persistent gender bias. Yet, not having an

---

[51]Ben Collins, Kevin Poulsen, and Spencer Ackerman, 'Russia' Facebook Fake News Could Have Reached 70 Million Americans', *Daily Beast*, 8 Aug. 2017.
[52]Thomas Rid, 'How Russia Pulled Off the Biggest Election Hack in U.S. History', *Esquire*, 20 Oct. 2016.
[53]Director of National Intelligence, *Background to Assessing Russian Activities and Intentions in Recent US Elections: The Analytical and Cyber Incident Attribution* (6 Jan. 2017).

effective counter to Russian information warfare and its convenient network of allies was a costly mistake.

From the strategic perspective, the benefit to Russia was in causing chaos in its target. This classic tactic of Russian disinformation campaigns continues to yield unforeseen benefits for Russia. These manipulation strategies engage on all fronts, seeking to achieve effects in situations where Russia has few advantages. Changing the direction of the US government, or weakening the new President before they even take office, is an enormous benefit to Russia in that it confuses policy toward Ukraine, delays any action against Syria for human rights violations, and allows Russia's operations to continue in a similar fashion against Germany and France during their 2017 elections.

Yet, overall, the greatest benefit was likely a bit more subtle; American alliances are fractured and confused. The US paradoxically offers military hardware to its allies while also threatening trade wars with these very same countries. If confusion was the goal, Russia succeeded to dramatic effect. Cyber espionage can have a clear and coercive effect, but it is rare, contingent on many factors, and depends on the lack of trust in the target on critical sources of information to achieve results. By preparing for massive global catastrophes that might be a cyber 9/11 scenario, observers miss the more persuasive and insidious impact of Russia's complex attack and dissemination strategies.

Ukraine: a case study of combined effects

The ongoing conflict in Ukraine offers a portrait of how Russia combines cyber coercion with other instruments of conventional power. This is also a critical case in that it goes beyond the US election hack with the use of information operations in support of conventional military operations. Specifically, cyber campaigns in Ukraine seek to disrupt and delegitimize the country as a means of isolating Kiev and demonstrating the futility of the Ukrainian state.

Russia seeks more than manipulation; it seeks domination. According to Giles, "unlike Georgia … Russia already enjoyed domination of Ukrainian cyberspace, including telecommunication companies, infrastructure, and overlapping networks."[54] This access allowed them to wage a more sophisticated coercive campaign. That said, there is no evidence to suggest the Russian campaign is that sophisticated. Rather, the digital domain played a supporting role to Russian proxy military operations and propaganda efforts. Outside of Russia's fait accompli seizure of Crimea, these operations have produced no concessions beyond producing a frozen conflict. In this

---

[54]Kier Giles, 'Putin's Troll Factories', *Chatham House* 71/4 (2015).

respect, cyber-combined coercion in Ukraine demonstrated that cyber options are restrained even in war. The disruptive campaign was a testing ground for new operations and no operation to date has proved decisive enough for the Ukrainians to back down to Russian aggression.[55]

In late 2013, activists set up a protest camp in Kiev's independence square (Maidan), calling for deeper European integration and an end to rampant corruption. The events escalated after over 100 protesters were killed in 48 hours by a special police unit, the Berkut, causing the protests to escalate and pro-Russian President Viktor Yanukovych to flee. By 27 February 2014, reports of Russian operatives and "local militias" in Crimea started to appear paving the way for a March referendum for Crimea to join Russia. In early May, two regions dominated by ethnic Russians, Donetsk and Luhansk, held referendums, declaring independence.

Paralleling the escalating crisis in late 2013, Ukrainian officials noted that "network vandalism had given way to a surge in cyber espionage, from which commercial cyber security companies developed a list of colorful names: RedOctober, MiniDuke, NetTraveler, and many more."[56] Koval claims that Russia conducted constant, low-level disruption campaigns against Ukraine using "botnet-driven" DDoS often "in retaliation for unpopular government initiatives."[57] Botnet-driven DDoS attacks involve overloading servers with content generated by hacked computers, usually without the knowledge of the user. Glib Pakharenko offers an insider's account of the cyberattacks during the Maidan protests:

> the cyber attacks began on 2 December 2013 when it was clear that protesters were not going to leave Maidan. Opposition websites were targeted by DDoS attacks, the majority of which came from commercial botnets employing BlackEngery and Dirt Jumper malware.... As Ukrainian opposition groups responded with their DDoS attacks, cyber-criminal organizations proactively reduced their use of the Ukrainian Internet Protocol (IP) space rerouting their malware communications through Internet Service Providers (ISP) in Belarus and Cyprus, which meant that, for the first item in years, Ukraine was not listed among the leading national purveyors of cybercrime.[58]

In early 2014, Ukrainian civilian and government networks were subject to a barrage of DDoS attacks. Cyber Berkut, a pro-Russian hacktivist group, a major proxy group with links to the Russian government, organized in the incidents. The threat actor took their name from the former special police unit disbanded in the wake of the Maidan protests, producing an illusion that pro-Moscow Ukrainians were rebelling against Kiev. CrowdStrike has

---

[55]Andy Greenberg, 'How an Entire Nation Became Russia's Test Lab for Cyberwar', *Wired*, 20 Jun. 2017.
[56]Nikolay Koval, 'Revolution Hacking', in Kenneth Geers (ed.), *Cyber War in Perspective: Russian Aggression against Ukraine* (Tallinn: NATO Cooperative Cyber Defence Center of Excellence 2015).
[57]Koval 'Revolutionary Hacking', 55.
[58]Koval 'Revolutionary Hacking', 50.

linked the group to the Russian government based on forensic data and parallels between messages put out by CyberBerkut and "messaging delivered by Russia-owned state media."[59] According to reporting by the firm,

> there are significant parallels between the current techniques employed by CyberBerkut and those used in previous conflicts associated with Russia, namely the conflict in Estonia in 2007. These techniques, leveraging Soviet-style deception, propaganda, and denial tactics, suggest a process in which the first iterations of online warfare implemented in Estonia are now being perfected in Ukraine.[60]

Cyber Berkut's actions ranged from disrupting mobile phone networks as a means of complicating Ukraine's response to the ongoing crisis to more complex, foreign disruptions designed to isolate and delegitimize Kiev.[61] In March 2014, Cyber Berkut claimed credit for a DDoS targeting three NATO websites. In October of that same year, the group was linked to a DDoS attack against German Ministry of Defense. In January 2015, a DDoS attack against the German Parliament and Chancellor Angela Merkel's websites was attributed to the group.

Over the course of 2014, Cyber Berkut also conducted prominent website defacements, placing narratives and symbols that matched Russian propaganda linking the Ukrainian conflict to fascism. In August 2014, the group hacked Polish websites, including the stock exchange, and defaced them with images of the Holocaust.[62] In November 2014, during Vice President Joe Biden's visit to Kiev, the group defaced several Ukrainian government websites with messages stating, "Joseph Biden is the fascists" master.'[63] In December 2014, the group hacked multiple electronic billboards in Kiev and replaced advertisements with video's showing graphic images of civilian casualties and portraying Ukrainian officials and anti-Russian activists as war criminals.

The most significant disruptive effort involved combining cyber espionage and disruption alongside propaganda to undermine the legitimacy of the Ukrainian election in 2014. According to Ian Gray, an analyst at Flashpoint, Russia seeks to achieve a low-cost disruption "by organizing a disinformation campaign attacking confidence in the election itself."[64] In May 2014, CyberBerkut "infiltrated Ukraine's central election computers and deleted key files, rendering the vote-tallying system inoperable. The next day, the hackers declared they had 'destroyed the computer network

---

[59]CrowdStrike, *2015 Global Threat Report* (2015).
[60]CrowdStrike, *2015 Global Threat Report*.
[61]Sam Masters, 'Ukraine Crisis: Telephone Networks are First Casualty of Conflict', *The Independent*, 25 Mar. 2014.
[62]Cory Bennett, 'Hackers breach the Warsaw Stock Exchange', *The Hill*, Oct. 2014.
[63]Vitaly Shevchenko, 'Ukraine Conflict: Hackers Take Sides in Virtual War', *BBC*, 20 Dec. 2014.
[64]Shaun Waterman, 'Russia Seeks to Discredit, Not Hack Election Results', *Cyberscoop*, 7 Nov. 2016.

infrastructure' for the election, spilling e-mails and other documents onto the web as proof."[65] Compounding the intrusion, the group installed malware that attempted to manipulate the results, showing a victory by ultranationalists, a key theme reinforced by broader Russian propaganda reflecting the Maidan as a Fascist revolution. According to Ukrainian cyber security experts, "preparation for such an attack does not happen overnight; based on our analysis of Internet Protocol (IP) activity, the attackers began their reconnaissance in mid-March 2014 – more than two months prior to the election."[66]

Another combined strategy on display in Ukraine was the use of false flag operations designed to not only hide attribution but also discredit the target, a tactic consistent with Soviet practices. False flags are a form of covert action designed to manipulate perception with deep historical roots. The term refers to making it seem as if an act was carried out under another nation's flag. A group attempts to conceal its involvement by creating the perception that a separate group carried out some act of sabotage, subversion, or physical attack. Under the handle Anonymous Ukraine, in March 2014, Russia released fabricated documents claiming to show evidence that the US Army Attaché was coordinating a series of false flag attacks designed to look like Russian Special Forces with the Ukrainian Army. In March 2015, CyberBerkut released documents said to be hacked from US defense contractors and Ukrainian government showing US plans to move weapons into Ukraine. Later that month, the group also released documents said to be hacked from the Ukrainian military showing that the government in Ukraine supplied weapons to the Islamic State. The group also released documents said to be hacked from the Soros Foundation showing that George Soros was pressuring American officials to provide lethal assistance to Ukraine.

In all cases, these false flag operations were picked up and broadcast through Russian media outlets and operatives on social media sites. Russia uses troll factories to shape how its public digests western media and distort unfavorable stories for foreign audiences.[67] For example, in March 2015, false flag operation citing evidence that the Ukrainian state-owned defense conglomerate Ukroboronprom collaborated with the Qatari government to supply surface-to-air missiles was reported on outlets such as *Sputnik International*, a state-controlled Russian media outlet.

This disinformation and delegitimizing campaign built on earlier network exploitation and cyber espionage. Access to Ukrainian information networks allowed them to spearfish Ukrainian officials. Hackers use typosquatting, registering a domain with just a misplaced letter, to spear phish

---

[65]Mark Clayton, 'Ukraine election narrowly avoided "wanton destruction" from hackers', *Christian Science Monitor* (17 Jun. 2014).
[66]Koval, 'Revolutionary Hacking', 60.
[67]Giles, 'Putin's Troll Factories'.

users accessing the website of Ukrainian President Petro Poroshenko.[68] Similar social engineering hacks were used as part the US presidential election hack, where the two groups, CozyBear and Fancy Bear, spearfished Democratic operatives at the DNC.[69] The same groups were also linked to spearfishing attacks on DC-area think tanks after the US presidential election.

As the Ukrainian crisis continued, Russia found new ways of combining cyber effects with irregular and conventional military operations. First, Russia employed traditional military operations to isolate information objectives. For example, in November 2014, Russian operatives sabotaged cables connecting the Crimean Peninsula to Ukraine.[70] Conventional operations helped isolate a target. Russia achieved similar effects in cyberspace. According to Glib Pakharenko:

> Russian signals intelligence (SIGINT) including cyber espionage, has allowed for very effective combat operations planning against the Ukrainian Army. Artillery fire can be adjusted based on location data gleaned from mobile phones and Wi-Fi networks. GPS signals can also be used to jam aerial drones. Ukrainian mobile traffic can be rerouted through Russian GSM infrastructure via a GSM signaling level (SS7) attack; in one case this was accomplished through malicious VLR/HLR updates that were not properly filtered. Russian Security Services also use the internet to recruit mercenaries.[71]

Second, Russia employed cyber methods to degrade Ukrainian military capabilities. In 2016, the cyber security firm Crowdstrike reported that Russia used an Android-based malware to infect apps Ukrainian units were using to compute the math required for targeting artillery. These infections enabled digital reconnaissance and helped Russian units geolocate Ukrainian artillery formations and preemptively strike them.[72] While there is some debate as to the effectiveness of this operation, with Crowdstrike altering the estimates from 80 percent effectiveness in targeting Ukraine to 15–20 percent, that Ukrainian artillery was using basic apps for targeting demonstrates the potential vulnerabilities that cyber operations can exploit in support of conventional military operations.

Third, Russia employed earlier cyber espionage campaigns to activate malware capable of degrading Ukrainian critical infrastructure. In October 2014, Sandworm used BlackEngery 3 to gain access to power plants and then insert KillDisk malware, a program similar to destructive systems used

---

[68]Patrick Tucker, 'The Same Culprits That Targeted US Election Boards Might Have Also Targeted Ukraine', *Defense One*, 3 Sep. 2016.
[69]Jeff Stone, 'Meet Fancy Bear and Cozy Bear, Russian Groups Blamed for DNC Hack', *Christian Science Monitor*, 15 Jun. 2016.
[70]Chris Baraniuk, 'Could Russian Submarines Cut off the Internet?', *BBC*, 26 Oct. 2015.
[71]Pakharenko, 55.
[72]Adam Myers, *Danger Close: Fancy Bear Tracking of Ukrainian Field Artillery Units* (CrowdStrike 22 Dec. 2016).

during the 2014 Ukrainian election.[73] Soon after the intrusion, some Ukrainian power plants went offline, though analysts still cannot draw a direct connection to KillDisk. Blackenergy 2015 and new CrashOverride 2016 are other critical infrastructure-targeting strands of malware that have been found in Ukrainian power plants.

Fourth, Russia integrated their operations with cyber disruption efforts paralleling broader disinformation campaigns. Russia weaponized social media to promote its narrative. Russian groups built redirects to steer users away from websites to Russian propaganda. For example, in 2015,

> cybercriminals helping spread pro-Russia messaging by artificially inflating video views and ratings on a popular video website. The campaign began with the infamous Angler exploit kit infecting victims with the Bedep trojan. Infected machines were then forced to browse sites to generate ad revenue, as well as, fraudulent traffic to a number of pro-Russia videos.[74]

Analysts linked the same malware to Russian cybercrime groups who used it to steal $45 million dollars from banks.[75]

In effect, Russia practices a new style of information operations designed "not to rebut but to obfuscate."[76] These operations rely

> on the fact that Western governments simply lack resources that would be required systematically to refute or debunk the huge number of stories put out, and on the Western media's professional obligation to report both sides of the story, thereby giving a veneer of legitimacy to Russian fabrications.[77]

Although Russian information and cyber operations reflect a degree of innovation not yet seen to such an extent in international campaigns, it is unclear if these strategies exhibit any novel utility. The media often seems to either ignore the implications of efficacy of cyber operations or present the issue from a purely partisan perspective (i.e., Russia did it but Trump's election is legitimate). These questions are too important to be investigated from a superficial perspective. The efficacy of cyber operations is critical and few have examined the issue empirically. Our case study here and the work of Kostyak and Zhukov demonstrate limited coercive utility through cyber espionage means.[78]

---

[73]Hultquist, *Sandworm team*.
[74]Rami Kogan, *Bedep Trojan Malware Spread by the Angler Exploit Kit gets Political* (Trustwave 29 Apr. 2015).
[75]Nick Biasini, 'Connecting the Dots Reveals Crimeware Shake-Up', *Talso*, 7 Jul. 2016.
[76]Nigel Inkster, 'Information Warfare and the US Presidential Election', *Survival* 58/5 (2016), 23–32.
[77]Inkster, 'Information Warfare'.
[78]Nadia Kostyak and Yuri Zhukov, 'Invisible Digital Front: Can Cyber Attacks Shape Battlefield Events?', *Journal of Conflict Resolution* (Forthcoming).

Conclusion

Russia prefers cyber disruptions that harass and sow discontent, which fail to coerce in a direct manner. The Russian approach to cyber strategy appears to be more about ambiguous signaling and amplifying propaganda than it does direct compellence. Russian cyber activities continue the Soviet approach to active measures, political warfare optimized to manipulate target populations and disrupt rivals from within. Due to these strategies, degradation efforts do not generate concessions in a manner similar to the cyber superpower, the US.

The Russian case offers insights into how states combine cyber operations with the military instruments of power or conventional information operations. Moscow tends to use cyberattacks in three waves: prior to the conflict to delegitimize and distract their rival, during the conflict to support combat operations, and after the initial fighting to create chaos that, consistent with active measures, undermines the legitimacy of the target state. However, Russian use of cyber operations during conflict does not appear to alter the outcomes or make concessions more likely. Of note, both the conflicts in Georgia and Ukraine have resulted in frozen conflicts, not decisive victories.

When Russia employs cyber coercion against Western rivals outside of its former Soviet space, these states tend to counter with diplomatic and economic instruments. These counters are consistent with a "tit-for-tat" logic, limited horizontal escalation designed to check Moscow and limit a dangerous conflict spiral. When cyber coercion is used against targets in the Baltics, the states counter by strengthening their ties to NATO and enhancing their domestic military.

Finally, Russian cyber strategy, in addition to reflecting tenets of Soviet-era active measures, focuses on soft targets, including civilian networks. These methods have largely been unsuccessful beyond the debatable example of the US Election hack in 2016. As opposed to the US, Russia tends to amplify propaganda with bots and troll farms rather than more traditional diplomatic coercion. To date, these operations have yet to produce concessions. In the end, Moscow acts more like a rogue state undermining the norm against targeting critical infrastructure than it does like a responsible actor in the digital domain.

Russia still has not unleashed the full potential of cyber operations against critical energy targets. They have not resorted to direct cyber violence, destroying infrastructure that results in immediate death such as blowing up a power plant, digitally sabotaging vital public infrastructure like sewers or water treatment, or hacking personal medical devices. Rather, Moscow's network intrusions on the battlefield in Ukraine indirectly helped military units increase their lethality.

Russia's aggressive, albeit unsuccessful, cyber operations threaten stability in cyberspace by targeting critical systems and illustrate how political warfare works in the digital domain. This case demonstrates the limits and logic of cyber strategies. While cyber operations do not produce significant effects on their own, they support other lines of effort including manipulating perception and sowing chaos in targeted populations.

Rather than herald a revolutionary break in the history of warfare, the employment of cyber operations between rivals can create strategic stability and reinforce traditional power dynamics. For Gartzke and Lindsay, there is a distinct stability–instability paradox in cyberspace.[79] The open architecture of the internet creates a unique vulnerability. If the target of coercion disconnects, they are no longer as vulnerable. Therefore, the aggressor has to operate either covertly or beneath a threshold to avoid retaliation. Furthermore, the aggressor knows that if it crosses that threshold, they risk a cross-domain response. A state could respond with economic sanctions, as seen in the Russia hack of US elections, or outright military force.

Since most forms of cyber strategy investigated here are optimized for covert action, they reflect a desire to signal resolve while keeping conflicts limited. Like covert action, cyber works in the shadows and can help rivals engage in tacit cooperation "to steer dangerous encounters to the backstage as a way to safeguard the external impression of their encounter as a limited conflict."[80] Plausible deniability in cases where attribution is fairly obvious (e. g., Russian incursions into Ukraine in 2014–15 or attacking the US election in 2016) works not necessarily to hide the identity of the attacker but rather to provide justification for the defender to moderate their response.

This signaling dynamic leaves us with the question of how do you compel a rival state if they do not know they have been breached in the first place? That is, how can covert action compel rivals? In 2016, the Obama Administration planted "cyber bombs" in Russian networks as a retaliation for the election hack.[81] Yet, if you do not signal the opposition that you have this deadly tripwire installed, how can you expect to affect their behavior and compel them to back down in their efforts to attack the American democratic process? Furthermore, if you send too explicit a signal, you give the target the opportunity to patch their network reducing your coercive leverage.

Despite the promise of quick wins in the digital domain, there are complex signaling dynamics in cyberspace that make producing concessions difficult. The future of cyber operations could reflect a drastically increased utility for

---

[79]Gartzke and Lindsay, 'Coercion through the Cyberspace'.
[80]Austin Carson, 'Facing Off and Saving Face: Covert Intervention and Escalation Management in the Korean War', *International Organization* 70/1 (2016), 105.
[81]Austin Carson, 'Obama Used Covert Retaliation in Response to Russian election Meddling. Here's Why', *Washington Post*, 29 Jun. 2017.

the efficacy of cyber operations. If planning is based on the current effective-ness of Russian operations, the best advice notes that their operations are restrained, generally fail to achieve results, and seek to limit escalation. Understanding this process of cyber strategies is a key task and results demonstrate more bluff and bluster than bending the will of the enemy.

Disclosure statement

No potential conflict of interest was reported by the authors.

Bibliography

BAE Systems, *The Snake Campaign* (Feb. 2014).

Baraniuk, Chris, 'Could Russian Submarines Cut off the Internet?', *BBC*, 26 Oct. 2015.

Beaufre, Andre, *An Introduction of Strategy* (London: Faber and Faber 1965 R.H. Barry translation).

Bennett, Cory, 'Hackers Breach the Warsaw Stock Exchange', *The Hill*, Oct. 2014.

Biasini, Nick, 'Connecting the Dots Reveals Crimeware Shake-Up', *Talso*, 7 Jul. 2016.

Borghard, Erica and Shawn Lonergan, 'The Logic of Coercion in Cyberspace', *Security Studies* 26/3 (2017), 452–81. doi:10.1080/09636412.2017.1306396

Carson, Austin, 'Facing off and Saving Face: Covert Intervention and Escalation Management in the Korean War', *International Organization* 70/1 (2016), 103–31. doi:10.1017/S0020818315000284

Carson, Austin, 'Obama Used Covert Retaliation in Response to Russian Election Meddling. Here's Why', *Washington Post*, 29 Jun. 2017.

Carson, Austin and Keren Yarhi-Milo, 'Covert Communication: The Intelligibility and Credibility of Signaling in Secret', *Security Studies* 26/1 (2017), 124–56. doi:10.1080/09636412.2017.1243921

Clayton, Mark, 'Ukraine Election Narrowly Avoided 'Wanton Destruction' from Hackers', *Christian Science Monitor* (17June2014).

Collins, Ben, Kevin Poulsen, and Spencer Ackerman, 'Russia' Facebook Fake News Could Have Reached 70 Million Americans', *Daily Beast*, 8 Aug. 2017.

CrowdStrike, *2015 Global Threat Report*, (2015).

Director of National Intelligence, *Background to Assessing Russian Activities and Intentions in Recent US Elections: The Analytical and Cyber Incident Attribution* (6 Jan. 2017).

Downs, George and David Rocke Downs, *Tacit Bargaining: Arms Races, Arms Control* (Ann Arbor: University of Michigan Press 1990).

Fearon, James D., 'Rationalist Expectations for War', *International Organization* 49/3 (1995), 379–414. doi:10.1017/S0020818300033324

Fearon, James D., 'Signaling Foreign Policy Interests Tying Hands versus Sinking Costs', *Journal of Conflict Resolution* 41/1 (1997), 68–90. doi:10.1177/0022002797041001004

Freedman, Lawrence, 'Strategic Studies and the Problem of Power', in Thomas Mahnken and Joseph A. Maiolo (eds.), *Strategic Studies: A Reader* (New York: Routledge 2008).

Gallagher, Sean, 'Seven Years of Malware Linked to Russian State-Backed Cyber Espionage', *Arstechnica*, 17 Sept. 2015.

Gartzke, Erik and Jon R. Lindsay, 'Coercion through the Cyberspace: The Stability-Instability Paradox Revisited', in Kelly Greenhill and Peter Krause (eds.), *The Power to Hurt in the Modern World* (Oxford: Oxford University Press 2017).

George F. Kennan on Organizing Political Warfare, 'History and Public Policy Program Digital Archive, Obtained and Contributed to CWIHP by A. Ross Johnson', *Cited in his book Radio Free Europe and Radio Liberty, Ch1 n4 – NARA release courtesy of Douglas Selvage. Redacted final draft of a memorandum dated May 4, 1948, and published with additional redactions as document 269, FRUS, Emergence of the Intelligence Establishment*, 30 Apr. 1948.

Gerasimov, 'The Value of Science in Prediction', in *Military-Industrial Kurier* (27Feb.2013).

Giles, Kier, 'Putin's Troll Factories', *Chatham House* 71/4 (2015).

Greenberg, Adam, 'Russia Hacked "Older" Republican Emails, FBI Director Says', *Wired*, 10 Jan. 2017.

Greenberg, Andy, 'How an Entire Nation Became Russia's Test Lab for Cyberwar', *Wired*, 20 Jun. 2017.

Hacquebord, Feike, *Operation Pawn Storm Ramps up Its Activities, Targets NATO, White House* (Trend Micro 16 Aug. 2015).

Healy, Patrick, David Sanger, and Maggie Haberman, 'Donald Trump Finds Improbable Ally in WikiLeaks', *New York Times*, 12 Oct. 2016.

Heuser, Beatrice, *The Evolution of Strategy: Thinking War from Antiquity to the Present* (New York: Cambridge University Press 2010).

Hulcoop, Adam, John Scott-Railton, Peter Tanchak, Matt Brooks, and Ron Deibert, *Tainted Leaks: Disinformation and Phishing with a Russian Nexus* (Citizen Labs May 2017).

Hultquist, John, *Sandworm Team and the Ukrainian Power Authority Attacks* (FireEye 7 Jan. 2016).

Inkster, Nigel, 'Information Warfare and the US Presidential Election', *Survival* 58/5 (2016), 23–32.

Isaac, Mike and Daisuke Wakabayashi, 'Russian Influence Reached 126 Million Through Facebook Alone', *New York Times*, 30 Oct. 2017.

Jensen, Benjamin, *Forging the Sword: Doctrinal Change in the U.S. Army* (Pal Alto: Stanford University Press 2016).

Kaspersky Labs, *Lazarus under the Hood* (25 Nov. 2017).

Kogan, Rami, *Bedep Trojan Malware Spread by the Angler Exploit Kit Gets Political* (Trustwave 29 Apr. 2015).

Kostyak, Nadia and Yuri Zhukov, 'Invisible Digital Front: Can Cyber Attacks Shape Battlefield Events?', *Journal of Conflict Resolution* (forthcoming).

Koval, Nikolay, 'Revolution Hacking', in Kenneth Geers (ed.), *Cyber War in Perspective: Russian Aggression against Ukraine* (Tallinn: NATO Cooperative Cyber Defence Center of Excellence 2015), 55–65.

Legum, Judd, 'Trump Mentioned WikiLeaks 164 Times in the Last Month of Election, Now Claims It Didn't Impact One Voter', *Think Progress*, 8 Jan. 2017.

Levy, Jack, 'Deterrence and Coercive Diplomacy: The Contributions of Alexander George', *Political Psychology* 29/4, 537–52.

Libicki, Martin, *Crisis and Escalation in Cyberspace* (Santa Monica: Rand Corporation 2012).

Lichtblau, Eric, 'Computer Systems Used by Clinton Campaign are Said to Be Hacked, Apparently by Russia', *New York Times*, 20 Jul. 2016.

Lindsay, Jon R., 'Stuxnet and the Limits of Cyber Warfare', *Security Studies* 22/3 (2013), 365–404. doi:10.1080/09636412.2013.816122

Lindsay, Jon R., 'The Impact of China on Cybersecurity: Fiction and Friction', *International Security* 39/3 (2014), 7–47.

Masters, Sam, 'Ukraine Crisis: Telephone Networks are First Casualty of Conflict', *The Independent*, 25 Mar. 2014.

MSS Global Threat Response, *Emerging Threat: Dragonfly/Energetic Bear – APT Group* (Symantec 30 Jun. 2014).

Myers, Adam, *Danger Close: Fancy Bear Tracking of Ukrainian Field Artillery Units* (CrowdStrike 22 Dec. 2016).

Osgood, Robert, *The Entangling Alliance* (Chicago: University of Chicago Press 1962).

Pape, Robert, *Bombing to Win: Air Power and Coercion in War* (Ithaca: Cornell University Press 1996).

Peters, Sarah, 'MiniDuke, CosmicDuke APT Group Likely Sponsored by Russia', in *Dark Reading* (17 Aug. 2015).

Polityuk, Pavel, *Ukraine Investigates Suspected Cyber-Attack on Kiev Power Grid* (Reuters 20 Dec. 2016).

Prince, Brian, '"Operation Armageddon" Cyber Espionage Campaign Aimed at Ukraine', *Security Week*, 28 Apr. 2015.

Rid, Thomas, 'How Russia Pulled off the Biggest Election Hack in U.S. History', *Esquire*, 20 Oct. 2016.

Rid, Thomas, 'Disinformation: A Primer in Russian Active Measures and Influence Campaigns', Hearings before the Select Committee on Intelligence, United States Senate, One Hundred Fifteenth Congress, 30 Mar. 2017.

Rupar, Aaron, 'Former FBI Agent Details How Trump and Russia Team up to Weaponize Fake News', *Think Progress*, 30 Mar. 2017.

Sanger, David, 'D.N.C. Says Russian Hackers Penetrated Its Files, Including Dossier on Donald Trump', *New York Times*, 14 Jun. 2016.

Sanger, David and Steven Erlanger, 'Suspicion Falls on Russia as Snake Cyberattacks Target Ukraine's Government', *The New York Times*, 9 Mar. 2014.

Sanger, David and Charles Savage, 'U.S. Says Russia Directed Hacks to Influence Elections', *New York Times*, 7 Oct. 2016.

Schelling, Thomas, *Strategy of Conflict* (Cambridge: Harvard University Press 1960).

Schelling, Thomas, *Arms and Influence* (New Haven: Yale University Press 1968).

Scott, James and Drew Spaniel, *Know Your Enemies 2.0* (Institute for Critical Infrastructure Technology 2016).

Sharp, Travis, 'Theorizing Cyber Coercion: The 2014 North Korea Operation against Sony', *Journal of Strategic Studies* (2017).

Shevchenko, Vitaly, 'Ukraine Conflict: Hackers Take Sides in Virtual War', *BBC*, 20 Dec. 2014.

Slayton, Rebecca, 'What Is the Cyber Offense-Defense Balance? Conceptions, Causes, and Assessments', *International Security* 41/3 (2016), 72–109.

Stone, Jeff, 'Meet Fancy Bear and Cozy Bear, Russian Groups Blamed for DNC Hack', *Christian Science Monitor*, 15 Jun. 2016.

Threat Intelligence, *APT28: A Window into Russia's Cyber Espionage Operations* (FireEye 27 Oct. 2014).

Tucker, Patrick, 'The Same Culprits That Targeted US Election Boards Might Have Also Targeted Ukraine', *Defense One*, 3 Sept. 2016.

Valeriano, Brandon, Ben Jensen, and Ryan Maness, *Cyber Strategy: The Changing Character of Cyber Power and Coercion* (New York: Oxford University Press 2018).

Waddell, Kaveh, 'Why Didn't Obama Reveal Intel about Russia's Influence on the Election?', *The Atlantic*, 11 Dec. 2016.

Waterman, Shaun, 'Russia Seeks to Discredit, Not Hack Election Results', *Cyberscoop*, 7 Nov. 2016.

Zetter, Kim, 'Russian Sandworm Hack Has Been Spying on Foreign Governments for Years', *Wired*, 14 Oct. 2012.

The political-military dynamic in the conduct of strategy

John Kiszely

ABSTRACT

A constructive and effective interaction between politicians and their military advisers is an essential element in the successful conduct of strategy. The author draws on his own experience at the political-military interface and the operational level to argue that the political-military relationship is inherently problematic and that it has become increasingly so in the twenty-first century in the United Kingdom, partly due to changes in the character of conflict, but also due to an erosion of trust between political decision-makers and their senior military advisers. The article concludes that certain approaches need to be taken to resolve these challenges.

It is self-evident that one of the keys to successful grand strategy – strategy at government or coalition level – is to be found at the political-military interface, and in the constructive and effective interaction between politicians – or, more correctly, government ministers – and their military advisers. Here I use the word 'strategy' in its meaning as a process – the process of balancing the ends to be achieved (as defined by policy) with the ways and means available, including the development and application of ways and means to achieve the ends.

There is nothing new in an assertion that the civil-military relationship is complex and can be problematic. Samuel Huntington in *The Soldier and the State* (1957)[1] and Morris Janowitz in *The Professional Soldier* (1960),[2] both writing primarily about the United States, drew attention to the complexities of the relationship and the potential for adverse outcomes, although their focus was mainly on the constitutional position of the military and the possibility of military intervention in politics. This issue was also central to Amos Perlmutter's *The Military and Politics in Modern Times: On Professionals,*

---

[1] Samuel Huntington, *The Soldier and the State: The Theory and Politics of Civil-Military Relations* (Cambridge, MA: Belknap Press of Harvard University 1957).
[2] Morris Janowitz, *The Professional Soldier: A Social and Political Portrait* (Glencoe, IL: Free Press 1960).

Praetorians and Revolutionary Soldiers (1977.)[3] Other authors, among them Peter Feaver and Richard Kohn,[4] focused on a perceived cultural 'civil-military gap'; and Eliot Cohen in *Supreme Command* (2002)[5] examined the tensions between military and civilian leaders in wartime. Hew Strachan, in *The Politics of the British Army* (1997)[6] and *(2013The Civil-Military 'Gap' in Britain (2013)*,[7] showed that the assertion that the British armed forces were both apolitical and unpolitical was largely a myth, and that they were no strangers to political intrigue or, from time to time, to exerting covert influence in the political process. In *The Direction of War: Contemporary Strategy in Historical Perspective* (2013),[8] Strachan highlighted a misperception about the civil-military relationship: '[w]e have tended to assume that the danger is a military coup d`etat, when the real danger for western democracies today is the failure to develop coherent strategy'.[9] More recently *The Report of the Iraq Inquiry* (2016)[10] (the Chilcot Report) included a number of comments on, and recommendations for, the civil-military relationship in the United Kingdom.

The scope of this article is restricted to one aspect of civil-military relations: the political-military dynamic in the conduct of strategy. The article focuses on the UK, but draws comparisons more widely. It argues that this dynamic is inherently problematic, that it has become increasingly so in the twenty-first century and that certain approaches need to be taken to resolve these challenges.

The political-military dynamic: inherently problematic

There is no shortage of historical examples to suggest that the political-military dynamic is an inherently problematic aspect of strategy. In the UK, the First World War saw continuous tension between the so-called 'brass hats' (the generals) and 'the frocks' (frock-coated ministers). This included the conflict between Prime Minister David Lloyd George and the head of the army, General Sir William Robertson, which resulted in Robertson's dismissal, and the even better-known battle between the First Lord of the Admiralty,

[3]Amos Perlmutter, *The Military and Politics in Modern Times: On Professionals, Praetorians, and Revolutionary Soldiers* (New Haven, CT: Yale University Press 1977).
[4]Peter Feaver and Richard Kohn (eds.), *Soldiers and Civilians: The Civil-Military Gap and American National Security* (Cambridge, MA: MIT Press 2001). See also Peter Feaver, *Armed Servants. Agency, Oversight and Civil-Military Relations* (Cambridge, MA: Harvard University Press 1991), and Thomas S Szayna et al (eds.), *The Civil-Military Gap in the United States: Does it Exist, Why, and Does It Matter?* (Santa Monica, CA: Rand Corporation 2007).
[5]Eliot Cohen, *Supreme Command: Soldiers, Statesmen, and Leadership in War* (London: Free Press 2003).
[6]Hew Strachan, *The Politics of the British Army* (Oxford: ClarendonPress 1997).
[7]Hew Strachan, 'The Civil-Military "Gap" in Britain', *Journal of Strategic Studies* 26/2 (2003), 43–63.
[8]Hew Strachan, *The Direction of War: A Study in Strategy* (Cambridge: Cambridge University Press 2013).
[9]Strachan, *The Direction of War*, 97.
[10]Sir John Chilcot, *The Report of the Iraq Inquiry: Report of a Committee of Privy* Counsellors. (London: Dandy Booksellers Ltd 2016).

Winston Churchill, and the head of the Royal Navy, Admiral of the Fleet Lord Fisher, over the Dardanelles strategy, which resulted in the resignation of both men. In the Second World War, Churchill, as Prime Minister, was again involved in a tense, at times turbulent relationship with his service chiefs.[11]

Not that such political-military tension is peculiarly British. There have, for example, been dramatic clashes in the United States between politicians and the military. Perhaps the most renowned is that between President Harry Truman and General Douglas Macarthur in 1951 over the conduct of the war in Korea. Less well known is the contempt in which President John Kennedy held his most senior military advisers after the Cuban missile crisis in 1962, writing, 'The first advice I`m going to give my successor is to watch the generals and to avoid the feeling that just because they were military men their opinion[s] on military matters were worth a damn'.[12] The result of the toxic relationship between his successor, Lyndon Johnson, and the Joint Chiefs of Staff is well told by H R McMaster.[13] More recently, US Defense Secretaries have clashed with military leaders – most famously Donald Rumsfeld before and during the campaign in Iraq from 2003–06. But he was far from alone in this. Indeed, his successor, the relatively mild-mannered Robert Gates, in his candid memoirs, entitled one of the chapters 'Waging War on the Pentagon', and observed drily that '[t]he relationship between senior military leaders and the civilian commander-in-chief – the president – is often a tense one'.[14]

Elsewhere, notable examples include the relationship between Israeli Prime Minister David Ben-Gurion and his generals in the 1948 Arab-Israeli War, or, for that matter, Prime Minister Golda Meir and hers in 1973. In France in the 1960s President Charles de Gaulle was faced with open revolt from some senior generals, and as recently as 2017 President Emmanuel Macron found it necessary to sack his army chief, General Pierre de Villiers. It is also worth recalling that in 2016, Turkey saw an attempted military coup against President Recip Erdogan, with a subsequent purge of the military, and that in 2017 it was military action which led to the removal from office of President Robert Mugabe in Zimbabwe.

A political-military gap?

Although these examples show that tensions between political leaders and their military advisers are widespread, they also suggest that the interaction

---

[11]See, for example, Alex Danchev and Daniel Todman (eds.), *War Diaries 1939–1945: Field Marshal Lord Alanbrooke* (London: Weidenfeld and Nicholson 2001), xv, xvii-xviii.
[12]Quoted by HR McMaster, *Dereliction of Duty: Lyndon Johnson, Robert McNamara, the Joint Chiefs of Staff, and the Lies that Led to Vietnam* (New York: HarperCollins 1997), 28.
[13]McMaster, *Dereliction of Duty.*
[14]Robert Gates, *Duty: Memoirs of a Secretary at War* (London: WH Allen 2015), 573.

between them depends on many factors, notably the society, culture and type of government. There is, therefore, a need to exercise caution when making generalisations on the subject across different countries and continents. There is also a need to balance examples of fractious political-military relationships with examples of political leaders having high degrees of trust in their senior military advisers. In the UK, in recent times, often-cited examples are Prime Minister Margaret Thatcher and her Chief of the Defence Staff (CDS), Admiral Sir Terence Lewin, during the 1982 Falklands Conflict,[15] and Prime Minister Tony Blair and his first CDS, General Sir Charles Guthrie, from 1998–2001.[16]

A number of factors combine to make the political-military dynamic at the grand strategic level particularly problematic. First, that level presents a challenging environment for the conduct of strategy. Grand strategy is the realm of complexity, ambiguity and uncertainty. Paradoxes abound. Decisions are usually made under pressure, against the clock and for high stakes. Strong personalities and big egos are commonplace. Tension – and not always creative tension – between political decision-makers and their military advisers is inherent in the process.

This is exacerbated by the fact that in Britain, as in most western democracies, political and military leaders may have very limited experience of each other's world. Political leaders are likely to have limited understanding of military ethos and culture, of inter-service rivalry and single-service agendas (let alone the military proclivity for jargon and acronyms), and it will not come easily for them to put themselves into the 'military mind'. For their part, senior military leaders may find themselves facing new challenges at the strategic level. For example, many senior commanders will probably be unused to having to defend their views under cross-examination. They are likely, at least initially, to have a limited understanding of the political ethos and culture, and are unlikely to be able to put themselves easily into the 'political mind'. This is particularly the case for military leaders who have not had previous experience of working at the political-military interface within the Ministry of Defence (MOD) – and not all jobs in the MOD are at the political-military interface, let alone involve direct, regular contact with ministers. But whatever their previous experience, service chiefs are likely to acquire some political skills pretty quickly (if only by osmosis) since outmanoeuvring politicians, civil servants and each other in the battle for resources is, for better or for worse, a large part of their job as they see it. Indeed, in his memoirs, Denis Healey, Defence Secretary from 1964–70, recalled, only partly in jest, that, 'I sometimes felt that I had learnt nothing

---

[15]See, for example, Charles Moore, *Margaret Thatcher: The Authorized Biography, Volume 1; Not For Turning* (London: Penguin 2013), 697.
[16]See, for example, Peter Oborne, 'An Officer and a Politician', *The Spectator* (27 May 2000). Available at http://archive.spectator.co.uk/article/27th-may-2000/21/an-officer-and-a-politician.

about politics until I met the Chiefs of Staff. Each felt his prime duty was to protect the interests and traditions of his service'.[17]

Thus, despite what they may say in public at the time, ministers and service chiefs often live in less than perfect harmony with each other. The bone of contention is usually that of resources, with service chiefs believing that ministers are denying them the wherewithal to do their jobs, and knowing that their own service expects them to fight for the necessary resources, whilst ministers suspect, often with good reason, that the underlying motive behind any military advice may be an increase in those resources. This is unfortunate since mutual trust between political and military leaders is a necessary foundation for good strategy, and that trust must be earned.

There are also likely to be very variable levels of understanding of strategy itself, both among and between government ministers and the senior military. Many ministers come to their appointments with no experience whatsoever of grand strategy and little understanding of it. It will probably be a matter of luck as to whether a cabinet contains many or few members with a high level of strategic understanding. The same may be true of the military. In the past, military chiefs have not always been selected on the basis of their strategic competence or their experience at the military-strategic level, either nationally or internationally.[18] Not all military leaders will have attended courses at higher-level military colleges, such as the UK's Royal College of Defence Studies. Indeed, in the UK such attendance by those who have become CDS has been the exception rather than the rule. Without a common understanding of strategy and its conduct, politicians and the military are likely to differ in their perceptions of its key elements, for example the centrality of balancing ends, ways and means, and the fact that strategy is not a one-off, linear, sequential process, but needs to be continuous, dynamic and iterative if it is to remain relevant to developing circumstances and opportunities.

In addition, perceptions on each side are likely to differ about their respective roles and responsibilities in the conduct of strategy and the formulation of policy. In the past, some in the military have viewed strategy in terms of its Greek origin as *strategos* – the business of the general – for example, the aforementioned General Robertson, who, according to Keith Jeffery, believed that 'war strategy should be left entirely to the military'.[19] Indeed, in Britain at the outset of the First World War there was some acceptance on the part of political leaders that the military would play the

---

[17]Denis Healey, *The Time of My Life: My Autobiography* (London: Penguin 1989), 263.

[18]See, for example, Bill Jackson and Dwin Bramall, *The Chiefs: The Story of the United Kingdom Chiefs of Staff* (London: Brasseys 1992), 126, 134, 136, 182–83, 267.

[19]Keith Jeffery, 'Sir Henry Hughes Wilson, baronet', in *Oxford Dictionary of National Biography* (Oxford: Oxford University Press 2004).

leading role in the conduct of war and strategy.[20] This was, however, to change, and change radically, when Prime Minister Herbert Asquith was replaced by David Lloyd George in 1916.[21] It was not just French President Georges Clemenceau who believed that war was too important to be left to the generals.[22]

Whilst there is now acceptance of the principle of civilian primacy throughout western democracies, there are widely differing opinions on the optimum approach to political control of the military. Huntington's favoured approach was what he termed 'objective control' where 'the healthiest and most effective form of civilian control of the military is that which maximises professionalism by isolating soldiers from politics, and giving them as free a hand as possible in military matters'.[23] His antithesis – 'subjective control' – involved tighter political control of the military and more detailed direction of military action. Unsurprisingly, militaries tend to favour objective control. Equally unsurprisingly, government ministers tend to have a different perception, favouring closer – often much closer – control. As Eliot Cohen argues persuasively, it was the lack of supervision, direction and intervention rather than a surfeit of it that led to serious errors in a number of campaigns, notably in Vietnam, but also in the First Gulf War.[24]

A complicating factor has been the doctrinal acceptance within NATO of the operational level of war – the level and link between strategy and tactics, concerned with the planning and conduct of campaigns. Militaries have tended to look at this arena as primarily military business, with some senior officers believing that it should be an exclusively military, politics-free preserve. Indeed, in the 2005 invasion of Iraq the US commander, General Tommy Franks, told Deputy Defence Secretary, Paul Wolfowitz, 'Keep Washington focused on politics and strategy. Leave me the hell alone to run the war'.[25] This showed an alarming ignorance of both strategy and the rights and duties of government ministers. All the more surprising was that in his memoirs,[26] Franks saw no need to question the wisdom of his remarks. The far-from-cordial political-military relationship was unhappily repeated in that between Franks's successor, Lieutenant General Ricardo Sanchez, and the political leader deployed to Iraq, Paul Bremer.[27] In the UK,

---

[20]'Perhaps epitomized by the appointment of Field Marshal the Earl Kitchener of Khartoum as Secretary of State for War.'

[21]Jackson and Bramall, *The Chiefs*, 85–86.

[22]Georges Clemenceau: 'War is too important to leave to soldiers', commonly paraphrased as 'War is too important to be left to the generals', quoted in John Hampden Jackson, *Clemenceau and the Third Republic* (London: Hodder and Stoughton 1948).

[23]Cohen, *Supreme Command*, 4–5 Huntington, The Soldier and the State, 83-85.

[24]Cohen, *Supreme Command*, 4.

[25]Emphasis in original. Tommy Franks, *American Soldier* (New York: Regan 2004), 440.

[26]Franks, *American Soldier*.

[27]Thomas E Ricks, *The Generals: American Military Command from World War II to Today* (New York: Penguin 2012), 411.

there may have been senior officers who have privately shared Franks's sentiments, but few, if any, since the Second World War who have stated their feelings so bluntly in public.

Finally, as well as differences in perception between politicians and the military over their roles and responsibilities in the conduct of strategy are differing – sometimes diametrically opposite – approaches to the process. For example, the starting point for the military in determining strategy tends to be to focus first on the end state to be achieved, then working backwards to the present position.[28] For the politician it tends to be the opposite: focusing first on the immediate action required, working forwards in response to events and opportunities. In terms of timing, the military want policy decisions made as early as possible, allowing maximum time for military planning and preparation. Ministers, however, want to take decisions as late as possible, thus keeping options open for as long as possible.[29] There are also likely to be differences of political and military opinion, over the very objective of a campaign (the ends), how it is conducted (the ways) and the resources to be employed (the means). Unless educated to the contrary, the military tend to see the objective in terms of achieving military victory – destroying the enemy – with military success equating to strategic success. For the politician (or, at least, the statesman), it is the quality of the subsequent peace that defines strategic success. As to the conduct of conflict, the military tend to see this as military business. For the politician it is political business, not least because how the conflict is fought directly impacts on the quality of the peace. In Philip Bobbitt`s words, 'We must choose which sort of war we will fight...[in order] to set the terms of the peace we want'.[30] When it comes to resources, the natural military desire is to use the maximum available, and, where possible, to follow the [General Colin] Powell doctrine of 1992, using overwhelming force to achieve a quick victory, within clear political objectives and with a clear exit strategy. This allows built-in redundancy to cope with the unexpected and with Clausewitzian friction ('the force that makes the apparently easy so difficult').[31] The politician, however, tends to favours the minimum necessary expenditure and seeks to expose military over-insurance and bear down on 'gold plating'. In terms of decision-making, the military preference is to offer one recommended course of action for quick decision. The

---

[28]John Kiszely, 'The British Army and Thinking About the Operational Level', in Jonathan Bailey, Richard Iron and Hew Strachan (eds.), *British Generals in Blair's Wars* (Farnham: Ashgate 2013), 119–130, 128. See, for example, Ministry of Defence, *Joint Doctrine Publication 5–00, Campaign Planning*, (London: MOD 2013), Section 2.

[29]Desmond Bowen, 'The Political-Military Relationship on Operations', in Jonathan Bailey, Richard Iron and Hew Strachan (eds.), *British Generals in Blair's Wars* (Farnham: Ashgate 2013), 273–280, 275.

[30]Philip Bobbitt, *The Shield of Achilles: War, Peace and the Course of History* (London: Penguin 2003), 780.

[31]Clausewitz, *On War*, edited and translated by Michael Howard and Peter Paret (Princeton: Princeton University Press 1976), 121.

political preference, and often insistence, is to be presented with a range of options for decision, even at the expense of time. These tensions can be a source of considerable frustration for both military and political leaders.

There is often divergence, too, in the political and military approach to risk. Popular mythology expects the military to be far more ready than politicians to undertake military action and, indeed, this can sometimes be the case,[32] in part a product of the 'can do' attitude in which the British armed forces rightly pride themselves. This attitude can, however, some-times cause the military to understate the difficulties and risks. The wise politician will also wish to look for underlying motives for such enthusiasm; it is not unknown for militaries to wish to demonstrate their own utility or the capability of some part of the military, particularly when defence cuts are looming. There were certainly suggestions that the Royal Navy's enthu-siasm for sending a task force to the Falkland Islands in 1982 was not totally unconnected with proposed reductions in the size of the navy.[33] Often, though, the risk appetite of military leaders will be less than that of politi-cians; it is, after all, military leaders who will have to handle the immediate consequences of military failure. Ministers can find such caution frustrating and see a need to spur the military into action. Churchill famously said, 'Why, you may take the most gallant sailor, the most intrepid airman, or the most audacious soldier, put them at a table together – what do you get? The sum of their fears'.[34] The same thought may have occurred to politicians listening to the 'catalogue of disasters that were paraded by military leaders' in the run-up to the first Gulf War in 1990.[35] This perception of military negativity was repeated during the campaign in Afghanistan in 2001; for example, Alastair Campbell, Prime Minister Tony Blair's director of commu-nications, noted in his diary that Admiral Sir Michael Boyce, the CDS, was 'as ever telling us what we couldn't do rather than what we could'.[36]

Before moving on, it is important to note the role of the civil service in the conduct of strategy. In the MOD, civilian officials are responsible not only for bureaucratic management, support and coordination with other government departments, and, along with their military colleagues, for managing the defence budget, procuring equipment and supplying the armed forces. They are also responsible, among other things, for providing advice to ministers on strategy and the political and economic impact of

---

[32]For example, in the decision to send a task force to the Falkland Islands in 1982. See Hugh Bicheno, *Razor's Edge: The Unofficial History of the Falklands War* (London: Weidenfeld and Nicholson 2006), 24.

[33]Max Hastings and Simon Jenkins, *The Battle for the Falklands* (London: Michael Joseph 1983), 62; Alastair Finlan, 'War Culture. The Royal Navy and the Falklands Conflict', in Stephen Badsey, Rob Havers and Mark Grove (eds.), *The Falklands Conflict Twenty Years On: Lessons For the Future* (London: Frank Cass 2005), 193–212, 201.

[34]Emphasis in original. Harold Macmillan, *The Blast of War, 1939-1945* (London: Macmillan 1967), 562.

[35]Bowen, 'The Political-Military Relationship on Operations', 279.

[36]Quoted by Theo Farrell, *Unwinnable: Britain's War in Afghanistan 2001–2014* (London: The Bodley Head 2017), 73.

military plans. They therefore have an essential role to play in any process of balancing ends, ways and means in the planning and conduct of military operations. Their relationship with both politicians and the military can have its tensions, not least because it is often their lot to have to explain why things cannot be done. A particular tension with the military in times of ongoing campaigns is that, in Hew Strachan's words, '[c]ivil servants see their primary function as serving their ministers and saving them from embarrassment, not winning the war in hand'.[37] It is civil servants who provide the continuity within the MOD; ministers and military officers come and go (the military, typically, on 2- to 3-year tours of duty; ministers often for shorter periods), while civilian officials typically spend their whole career in the department. They represent the corporate memory and, compared to both politicians and the military, tend to take the long-term view. Politically neutral, civil servants in the MOD 'hold the ring' for the three service tribes but are not without their own agenda, and in that sense are, themselves, the fourth tribe.

The political-military dynamic today

There are a number of factors which have contributed to making the political-military dynamic in the conduct of strategy even more problematic since the end of the Cold War. The majority of these relate to the changing character of conflict and the prevalence of counter-insurgencies, stabilisation operations and conflicts involving non-state, or ostensibly non-state, actors (e.g., disguised Russian soldiers in Ukraine in 2014–15). Such campaigns present particular challenges to strategists and policy-makers alike. First, the levels of complexity, ambiguity, confusion and uncertainty are easily underestimated. There is a blurring of the difference between peace and war; defeat and victory are often meaningless terms; success and failure become comparative rather than absolute measurements. Conflicts such as counter-insurgency are intensely political affairs: 'twenty per cent military, eighty per cent political', according to French theorist David Galula, writing in the 1960s.[38] Today, the political nature is, arguably, even more heavily weighted. Ubiquitous media and the power of social media mean that military decisions and action are immediately subject to public scrutiny and even decisions taken at the lowest level can have strategic consequences – epitomised by the idea of the 'strategic corporal'.[39] The

[37]Hew Strachan, 'Conclusions', in Jonathan Bailey, Richard Iron and Hew Strachan (eds.), *British Generals in Blair's Wars* (Farnham: Ashgate 2013), 327–346, 341. See also Sir Mike Jackson, *Soldier: The Autobiography of General Sir Mike Jackson* (London: Corgi 2008), 358–359.
[38]David Galula, *Counterinsurgency: Theory and Practice* (Westport: Prager 2006), 63.
[39]General Charles Krulak, 'The Strategic Corporal. Leadership in the Three Block War', *Marine Corps Gazette* 83/1 (1999).

battleground is the mind – the minds of the indigenous population, and of regional and world opinion; shaping the narrative is central to success. In such circumstances, differences in perception and approach between politicians and military advisers can become sharper. For example, former Foreign Secretary Douglas Hurd observed that:

> British generals are always pressing for precision from politicians. But politicians live in a confused world; politicians will do their best to provide precision but must fail somewhat. Hence the military must share some of the burden of that confusion and be asked to make the best of a confused situation or a bad job. But the military are not trained to work with such a lack of precision.[40]

Second, with the speed of modern communications, politicians, understandably, want advice and answers quickly. Yet in planning these complex campaigns there is a danger of jumping to conclusions before the necessary in-depth analysis has taken place.[41] Easily underestimated is the time that such analysis takes. For example, the Iraq Inquiry criticised the sub-optimal 'appreciation of the theatre of operations, including the political, cultural and ethnic background, and the state of society, the economy and infrastructure'.[42] An almost identical point was made by a respected analyst about the British deployment to Helmand province, Afghanistan in 2006: 'the amount and quality of intelligence acquired in the short timeframe preceding the UK's deployment did not allow for a thorough appreciation of the complex tribal, social and political dynamics of the province'.[43] In both cases failures were greatly exacerbated by the pressure of time, resulting in corners being cut and analysis rushed.[44] Future deployments are unlikely to be any less demanding in this respect.

Third, since modern campaigns are invariably more than purely military affairs, and involve other, non-military lines of operations, such as diplomatic, economic and social, they require the commitment of departments across government and an integrated, comprehensive approach. This requirement is underlined by the fact that strategic problems in these campaigns are often what is termed 'wicked' – that is to say, complex, intractable interconnected problems where the solution to one problem creates or exacerbates problems elsewhere. For example, the decision in 2005 to follow a poppy-eradication programme in Afghanistan so alienated

---

[40]Quoted in Christopher Elliott, *High Command: British Military Leadership in the Iraq and Afghanistan Wars* (London: Hurst & Company 2015), 193.

[41]Royal College of Defence Studies, *Getting Strategy Right (Enough)* (London: MOD 2017), 55.

[42]*Report of the Iraq Inquiry*, (London: HMSO 2016), Executive Summary, 134.

[43]Valentina Soria, 'Flawed "Comprehensiveness": The Joint Plan for Helmand', in Michael Clarke (ed.), *The Afghan Papers. Committing Britain to War in Helmand, 2005–06* (Royal United Services Institute 2011), 30–48, 37.

[44]See Jack Fairweather, *The Good War: The Battle for Afghanistan 2006–2014* (London: Jonathan Cape 2014), 152.

the local population that the challenge for the security forces was greatly increased: '[i]n many ways, it gave them a cause'.[45] Yet a comprehensive, cross-government approach, so often called for by the military,[46] has proven highly problematic. Top-down political direction has often been lacking, and loose coordination has proved a poor substitute. Prime Ministers have found themselves, perhaps unsurprisingly, too busy to take personal charge of campaigns and unwilling to appoint a deputy to do so. Tony Blair created the Ad Hoc Group on Iraq in 2002 and the Iraq Planning Unit a year later. But as the Iraq Inquiry commented, '[n]either body carried sufficient authority to establish a unified planning process across the four principal departments involved…or between military and civilian planners', and that 'Mr Blair did not establish clear Ministerial oversight of post-conflict strategy, planning and preparation'.[47] Furthermore, the Inquiry observed that, following a visit to Iraq in June 2003, Blair told ministers that the British government should return to 'a war footing' to avoid 'losing the peace in Iraq', but that there were 'no indications that Mr Blair's direction led to any substantive changes in the UK`s reconstruction effort'.[48] The Prime Minister`s authority was, in the words of one commentator, 'unable to overcome bureaucratic inertia and conflicting ministerial and departmental agendas'.[49] One impact of this lack of direction and coordination was that at the tactical level in theatre, according to General Sir Richard Dannatt, Chief of the General Staff 2006–09, 'internecine squabbling over roles, resources and responsibilities dangerously damaged the combined effects we were trying to achieve'.[50] Similar civil-military friction was to characterise the UK's deployment into Helmand province in Afghanistan in 2006.[51]

Fourth, the fact that campaigns today almost always involve a coalition of nations – and often a large number of them – adds yet a further dimension of complexity to strategy and a further potential source of political-military friction. Politically, coalitions are attractive – they provide, for example, increased political clout, perceived legitimacy and shared cost – and, in

---

[45]General Sir Nicholas Houghton, evidence to the House of Commons Defence Committee, 6 July 2011, https://publications.parliament.uk/pa/cm201012/cmselect/cmdfence/554/11051101.htm.

[46]See, for example, Ben Barry, *Harsh Lessons: Iraq, Afghanistan and the Changing Character of War* (Abingdon: Routledge 2017), 65–71; General Sir Richard Dannatt, *Leading from the Front: The Autobiography* (London: Bantam Press 2010), 296; Andrew Mackay and Steve Tatham, *Behavioural Conflict: Why Understanding People and Their Motivations Will Prove Decisive in Future Conflict* (Saffron Walden: Military Studies Press 2011), 137.

[47]*Report of the Iraq Inquiry*, Summary, 81 and 86.

[48]*Report of the Iraq Inquiry*, Section 10.4, 535.

[49]Barry, *Harsh Lessons*, 49.

[50]Dannatt, *Leading from the Front*, 296. See also Jackson, *Soldier*, 418; and Justin Maciejewski, '"Best Effort": Operation Sinbad and the Iraq Campaign', in Jonathan Bailey, Richard Iron and Hew Strachan (eds.), *British Generals in Blair's Wars* (Farnham: Ashgate 2013), 157–174, 164.

[51]Soria, 'Flawed "Comprehensiveness"', 43; John McColl, 'Modern Campaigning: From a Practitioner's Perspective' in Jonathan Bailey, Richard Iron and Hew Strachan (eds.), *British Generals in Blair's Wars* (Farnham: Ashgate 2013), 109–118, 111–112.

general, the larger the coalition the better. But militarily, coalitions represent Clausewitzian friction – for example, in terms of interoperability, operational caveats, and differing rules of engagement – and, in general, the friction increases in direct proportion to the number of members.[52] Coalition operations are, by their very nature, intensely political affairs, with the need for compromise in both political objectives and the ways and means of pursuing them. Indeed, each coalition member will have its own, often competing, national agenda. To reach the necessary consensus, political direction to the military can be expressed in deliberately ambiguous, blurred language. But ambiguity and the resulting misunderstandings can directly contribute to disaster – for example, the so-called 'safe areas' in Bosnia in 1995.[53] Moreover, few decisions facing senior military commanders in multinational campaigns do not have a significantly political dimension, with implications for political involvement and control.

This leads to the fifth factor: the appropriate degree of civilian control – a subject which has proved to be a contentious political-military issue in campaigns in the twenty-first century. For example when, during the campaign in Afghanistan, British troops were deployed to the north of Helmand province the military viewed the deployment as just 'a change of tactics'.[54] For ministers, it was a major strategic decision, and one on which a number felt that they had not been properly consulted.[55] There was also political concern that important decisions were being left to tactical-level commanders in theatre. Indeed, when Defence Secretary John Hutton visited the British brigade in Helmand in 2009, he perceived that 'strategy was largely left to the military'.[56] Such delegation could be held to be in line with the British military doctrine of mission command which advocates giving maximum flexibility and discretion to subordinate commanders, but this was taking place to a questionable degree and at the expense of consistency. Every 6 months, a new brigade deployed to Afghanistan, its commander choosing how he would approach the campaign. One might choose a highly kinetic warfighting approach, his successor changing to one focused on the needs of the population; 6 months later the approach would change again. When they departed, '[e]very British brigade left Helmand declaring that some kind of turning point had been achieved' during their tour.[57] The MOD's director of operations from 2003–06, Lieutenant General Sir Robert

---

[52]John Kiszely, *Coalition Command in Contemporary Operations* (London: Royal United Services Institute 2008), 2–3.

[53]Shashi Tharoor, 'Should UN Peacekeeping Go "Back To Basics"?', *Survival* 37/4 (1995), 52–64, 60.

[54]General Sir David Richards, evidence to the House of Commons Defence Committee, 11 May 2011, https://publications.parliament.uk/pa/cm201012/cmselect/cmdfence/554/11051102.htm.

[55]Michael Clarke, 'The Helmand Decision', in Michael Clarke (ed.), *The Afghan Papers: Committing Britain to War in Helmand, 2005–06* (London: Royal United Services Institute 2011), 5–29, 21.

[56]Farrell, *Unwinnable*, 272.

[57]Farrell, *Unwinnable*, 367. See also Frank Ledwidge, *Losing Small Wars: British Military Failure in Iraq and Afghanistan* (New Haven, CT: Yale University Press 2011), 34, 85.

Fry, later commented, 'there emerged a culture where there was just too much deference to the commander on the ground'.[58] One seasoned observer and regular visitor to Afghanistan, Mark Urban, put it more bluntly. 'It was', he said, 'mission command gone bonkers'.[59]

Sixth, a further facet of contemporary campaigns that can cause friction between political leaders and their military advisers is that counterinsurgencies and conflicts with non-state actors tend, in practice, to be lengthy and unpredictable affairs – often much more so than original estimates. Whereas the military fully expect that the plan will change after the first encounter, 'the politician who has spelled out his expectations will regard it as close to criminally irresponsible and fear charges of incompetence'.[60] In addition, the military are likely to be arguing for the campaign to be extended until the necessary conditions in theatre are achieved. At the forefront of the politician's mind, though, will be the domestic political impact, particularly as measured by the opinion polls. Where the campaign is becoming unpopular, for example as a result of unexpectedly high casualties, the political decision is likely to favour a time-limited conclusion, with less regard for progress in theatre – often to the immense frustration of the military.[61]

But in addition to the factors relating to the changing character of conflict which have made the political-military dynamic even more problematic are a group of factors relating more to the practitioners of strategy themselves. The first of these concerns their experience. The direct military experience of senior British politicians (as of senior politicians in many countries) is much less than a generation ago. For example, at the time of the Falklands conflict in 1982, the British cabinet included six members with previous military service; now, in October 2017, the British cabinet has none. Nor has any Defence Secretary since 1992 had previous military service, and most came to the job with little knowledge of the military. Des Browne, appointed Defence Secretary in 2006 – a critical point in the campaigns in both Iraq and Afghanistan – freely admitted, 'I came to the job with no experience of military matters'.[62] Sherard Cowper-Coles, Britain's ambassador in Kabul from 2007–10, recalls suggesting to a visiting minister that he might want to question an MOD proposal to deploy Tornado aircraft to Afghanistan. 'His reply illustrated all the difficulties of civilian politicians with no military expertise assessing military advice. "Sherard", he said, "I don't know the difference between a tornado and a torpedo. I can't possibly question the Chief of the Defence Staff on this"'.[63] On another occasion a

---

[58]Quoted by Elliott, *High Command*, 177.
[59]Mark Urban, at RUSI conference 11 September 2017.
[60]Bowen, 'The Political-Military Relationship on Operations', 275.
[61]See, for example, Maciejewski, '"Best Effort"', 160.
[62]Quoted by Elliott, *High Command*, 43.
[63]Sherard Cowper-Coles, *Cables from Kabul: The Inside Story of the West's Afghanistan Campaign* (London: HarperCollins 2011), 282.

minister asked Cowper-Coles to remind him of the difference between a battalion and a brigade.[64]

In such circumstances it is easier for the military, for better or for worse, to influence policy. For example, the substantial size of Britain's contribution to the invasion of Iraq in 2003 was strongly influenced by the military's keenness to be involved on a large scale. In part this was due to a wish for the UK to be in a position to influence alliance policy, to be seen as a staunch ally by the United States and to cement the relationship with the American military, but it was also due to a strong desire to demonstrate the utility and value of the armed forces and justify the substantial resources invested in them. Moreover, within the armed forces, and for the same reason, each service was seeking to maximise its role. For example, the army pressed for, and achieved, a force package which included an armoured division, with a number of senior officers strongly advocating, although failing to achieve, the inclusion of the British-led headquarters of NATO's Allied Rapid Reaction Corps, which was thought to be vulnerable to defence cuts[65] – an argument summed up in the phrase, 'use it or lose it'. There was also significant senior military influence in the decision in 2005 to pivot military involvement from Iraq to Afghanistan, including a strong desire to be involved in 'a good war' in which British forces could excel, following unhappy experiences in Iraq.[66] Once in Afghanistan, according to Cowper-Coles, 'the British Army [went] to Helmand not to defeat the Taliban, but to defeat the British Treasury, the Royal Navy and the Royal Air Force [in the battle for funding]'.[67]

Yet if politicians can be said to lack military experience and understanding, so the changing character of conflict can bring into question the relevance of the experience of senior military leaders and their expertise as advisers.[68] For example, at the time of the Vietnam War, the US service chiefs of staff all had considerable warfighting experience, but none of the type of war on which they were embarking. John F Kennedy's caustic comment, quoted earlier, on his service chiefs may not have been wide of the mark. Similarly, as Western militaries embarked on counter-insurgency campaigns in Iraq and Afghanistan in the first decade of the twenty-first century, few service chiefs in coalition nations had direct experience of such campaigns. Ironically, almost all senior British army officers had experience

---

[64]Ibid.
[65]Bowen, 'The Political-Military Relationship on Operations', 277.
[66]Nick Beadle, 'Afghanistan and the Context of Iraq', in Michael Clarke (ed.), The Afghan Papers: Committing Britain to War in Helmand 2005–06 (Abingdon: Routledge 2011), 74–75; Ledwidge, Losing Small Wars, 58.
[67]Quoted by Elliott, High Command, 21.
[68]See John Kiszely, 'Post-Modern Challenges for Modern Warriors', in Patrick Cronin (ed.), The Impenetrable Fog of War: Reflections on Modern Warfare and Strategic Surprise (Westport: Prager 2008), 129–152.

of counter-insurgency – in Northern Ireland – but many were apt to over-rely on that single campaign in planning and conducting campaigns in what were very different circumstances, and, according to General Lord Richards, CDS from 2010–13, slow 'to think in new and innovative ways'.[69] Looking ahead, although many of today's senior military officers have considerable experience of counter-insurgency campaigns, none have any experience of what could conceivably be the next campaign: interstate, 'regular' warfare, or, indeed, hybrid warfare.

The second factor relating to the practitioners of strategy which can affect the political-military dynamic is the paradoxical approach often taken by the military to ongoing operations. On the one hand, there is considerable military frustration that during a major campaign in which servicemen and -women are fighting and dying, it appears to be 'business as usual' in Whitehall, with no sense that the country is at war.[70] Indeed, General Richards admitted that when he was Chief of the General Staff (2009–10) he had 'got pretty stroppy' with those who denied that Britain was engaged in a war.[71] Yet at the same time, the military often acts as if there were higher priorities, for example resisting pressure to extend the 6-month operational tour length for commanders and key personnel[72] and showing a reluctance to invest in ongoing operations at the expense of the long-term equipment programme. In the latter case the Iraq Inquiry noted that the Army found difficulty in funding an urgent operational requirement for up-armoured patrol vehicles in order, it suggested, to protect funding for a new range of armoured fighting vehicles, the Future Rapid Effects System, then under development for use in regular warfare. '[I]t appears that the longer-term focus of the Executive Committee of the Army Board on the Future Rapid Effects System programme inhibited it from addressing the more immediate issue related to [an up-armoured patrol vehicle]'.[73] The army saw things differently, pointing to the Treasury's duty to fund urgent operational requirements. Nor is this paradox peculiarly British; Robert Gates expressed frustration about very similar attitudes in the Pentagon during his time as Defense Secretary.[74]

The third and final factor pertaining to the practitioners is probably the most significant one. There is nothing new about service chiefs using media

---

[69]General David Richards, *Taking Command: The Autobiography* (London: Headline Publishing 2014), 78.
[70]Dannatt, *Leading from the Front*, 248.
[71]Richards, *Taking Command*, 294.
[72]Chris Brown, 'Multinational Command in Afghanistan – 2006: NATO at the Crossroads', in Jonathan Bailey, Richard Iron and Hew Strachan (eds.), *British Generals in Blair's Wars* (Farnham: Ashgate 2013), 217–224, 217; Kiszely, *Coalition Command in Contemporary Operations*, 19; UK Parliament, *House of Commons Defence Committee Thirteenth Special Report* (Oct 2007), paragraphs 50–51, and Government Response, paragraph 15.
[73]*Report of the Iraq Inquiry*, Summary, 127. See also Jack Fairweather, *A War of Choice: The British in Iraq 2003–2009* (London: Jonathan Cape 2011), 255–256.
[74]Gates, *Duty*, 117–126.

contacts to try and advance the cause of their service. Indeed, Michael Heseltine recalled that when he was Defence Secretary (1983–86) 'each of the services had an umbilical link…with the defence correspondents of the major newspapers'.[75] But in recent years the degree to which senior military officers have sought to influence political decisions by leveraging public support for their cause has opened them to accusations of straying into the political arena. Many of them would see it as no more their duty to fight on behalf of their service, and a dereliction of that duty if they failed to do so.[76] Thus, particularly when defence cuts are in prospect, senior military officers are busy giving journalists off-the-record briefings or suggesting to international military colleagues that they might voice concerns, either directly or through their political masters.[77] Such action can stretch to wider policy. For example, James de Waal asserts that the decision to make a major ground force contribution to the invasion of Iraq in 2003 was made 'primarily because politicians feared they would have problems with the British army if it were left out, and that these problems would find their way into the media'.[78] Similarly, he suggests that in 2009, 'Downing Street was not convinced of the military need to send reinforcements to Afghanistan, but agreed to do so because it wanted to prevent hostile press briefings by the military'.[79] Furthermore, there is evidence to support de Waal's assertion that during the Iraq and Afghanistan campaigns, minsters were 'apprehensive of the close relationship between the armed forces and the media and were therefore reluctant to challenge military opinion'.[80] He also perceives that the civil service voice in the conduct of strategy during these campaigns was distinctly muted since officials did not see the detail of military operations as part of their business.[81] In 2011 Prime Minister David Cameron reportedly became highly irritated by statements to the media by senior officers and told the CDS, General Richards, 'You do the fighting and I'll do the talking'.[82] Additionally, a number of recent service chiefs, very shortly after retirement, have written autobiographies, including in them advice given to ministers.[83] All of this has, unsurprisingly, contributed to eroding the trust of politicians in their military advisers – trust that is at the foundation of good strategy.

[75]Michael Heseltine, *Life in the Jungle: My Autobiography* (London: Hodder and Stoughton 2000), 262.
[76]See, for example, Dannatt, *Leading from the Front*, 245–251.
[77]See, for example, Deborah Haynes, 'US military officers raise fears over Royal Marine cuts', *The Times* (25 October 2017).
[78]James de Waal, *Depending on the Right People: British Political-Military Relations, 2001–10* (London: Chatham House 2013), vi.
[79]de Waal, *Depending on the Right People*, vi.
[80]de Waal, *Depending on the Right People*, vi.
[81]de Waal, *Depending on the Right People*, 25.
[82]Alex Massie, 'You Do the Fighting, I'll Do the Talking', *The Spectator* (20 June 2011).
[83]General Sir Richard Dannatt, *Boots on the Ground: Britain and her Army since 1945* (Profile Books 2016); Jackson, *Soldier*, Richards, *Taking Command*.

Mitigating the increased challenges

There are, however, a number of recent developments that mitigate these increased challenges to the political-military dynamic. Most obviously there have been a number of structural changes. The formation in 2010 of the National Security Council was a major step forward in the coordination of the response to national security threats and the formulation of national strategy, even if, in practice, it has tended to focus more on the former than the latter.[84] Significantly, military representation on the Council was restricted to the CDS, excluding the single-service chiefs of staff.

There have been further, if less obvious, developments. While the campaigns in Iraq and Afghanistan were still underway there was belated recognition in Whitehall that improvement was required in the conduct of strategy and in the interaction between political decision-makers and their advisers, both military and civilian. In December 2009 the CDS, Air Chief Marshal Sir Jock Stirrup, devoted his entire annual lecture at the Royal United Services Institute to the need to acquire 'the habit of thinking strategically'.[85] At around the same time the Royal College of Defence Studies refocused its syllabus more on to the conduct of strategy, to supplement its general study of the subject. In 2012 the MOD produced a new doctrinal publication, 'Organising Defence's Contribution to National Strategy'.[86] Additionally, a CDS's Strategic Advisory Panel, including external members from academia, was set up along with an annual CDS's Strategy Forum run in partnership with the Changing Character of War Centre at Oxford University. Further action followed the publication in 2016 of the Iraq Inquiry report. The Defence Secretary, Sir Michael Fallon, directed his department to 'embed the lessons of Chilcot in our DNA'[87] and set up a team under a senior official to extract lessons from the report and formalise their implementation.[88] At the same time he also expressed his determination to build within the MOD 'a culture of reasonable challenge' in which individuals felt empowered to challenge the views of their superior. Indeed, his department published detailed guidance on how this was to be carried out.[89] MOD has since produce a further three important publications aimed at improving strategy.[90] It remains to be seen, of course, whether the momentum of this action sustains.

---

[84]Joe Devanny and Joe Harris, *The National Security Council: National Security at the Centre of Government* (London: Institute for Government 2014), 30–32.

[85]CDS annual lecture at Royal United Services Institute (3 December 2009). Available at http://www.rusi.org/cdslectures.

[86]Ministry of Defence, *Organising Defence's Contribution to National Strategy* (London: MOD 2012).

[87]Quoted in Royal College of Defence Studies, *Getting Strategy Right (Enough)*, v.

[88]Martha Gill, 'Michael Fallon Sets Up Team To Trawl For Chilcot's Buried Recommendations', *Huffington Post* (19 July 2016).

[89]Quoted in Royal College of Defence Studies, *Getting Strategy Right (Enough)*, Annex A, 19.

[90]Ministry of Defence, *Making Better Strategy* (London: MOD, 2016), Ministry of Defence, *JDP 04. Understanding and Decision Making* (London: MOD 2016), and Royal College of Defence Studies, *Getting Strategy Right (Enough)*.

There have been other factors, too, that have helped to improve the political-military dynamic in the conduct of strategy. Participation in complex, counter-insurgency campaigns in Iraq and Afghanistan over a number of years has given greater understanding of the political-military issues, both to those deployed in theatre and to those involved in campaign planning and management in Whitehall and at the Permanent Joint Headquarters. Such lessons extend to the tactical level. For example, a young platoon commander in Afghanistan, Emile Simpson, recalled that:

> the mission, which I experienced as an infantry officer in southern Afghanistan, became indistinguishable from local politics. Given the need to tackle all the problems that stoked insurgency – poor governance, corruption, land rights, ethnic prejudice – it could not have been anything else.[91]

As such officers become more senior, their experience will reap benefits in the conduct of strategy at higher levels. Similarly, there are now a number of ex-military Members of Parliament with operational experience in Iraq and Afghanistan, some of whom already hold junior ministerial appointments.

Finally, the MOD has instructed the Permanent Joint Headquarters to provide 'politically aware military advice'.[92] If this is taken to mean military advice that takes into account political factors, it will be conducive to good strategy. If, however, it is aimed at (and succeeds in) discouraging advice that might present problems for ministers (e.g., in having to admit that they ignored military advice), it will be actively conducive to bad strategy.

Keys to success in the political-military dynamic

There are a number of keys to success in the political-military dynamic. First and foremost, is the requirement for a better understanding of strategy and of what Clausewitz termed 'the interplay of war and politics'. In particular, it requires an understanding and acceptance by all in the military that 'at the highest level, the art of war turns into policy',[93] and that '[a] commander-in-chief…must be familiar with the higher affairs of state and its innate policies; he must know current issues, questions under consideration, the leading personalities, and be able to form sound judgment'.[94] Likewise, it behoves political leaders to accept another of Clausewitz's dicta:

---

[91]Emile Simpson, 'America Must Leave the Defeat of Isis to Local States', *Financial Times* (11 September 2014).
[92]PJHQ, 'Our Organisation and Responsibilities'. Available at https://wgovernment/government/groups/the-permanent-joint-headquarters.
[93]Clausewitz, *On War*, 607.
[94]Clausewitz, *On War*, 146.

In the same way as a man who has not fully mastered a foreign language fails to express himself correctly, so statesman often issue orders that defeat the purpose they are meant to serve. Time and again that has happened, which demonstrates that a certain grasp of military affairs is vital for those in charge of general policy.[95]

One of the foundations of better strategic understanding, for both military and political leaders, is education. And since both political and military leaders will be far too busy when they approach, let alone fill, the top appointments, this education must take place earlier in their careers. Although much can come from self-education – from serious study of the subject – an important element is formal institutional education. In the UK such education is already provided for a small number of military officers at the Royal College of Defence Studies, and a recent initiative has been to provide a few places for parliamentarians to attend a term of the course. The number should be increased, and greater care made by the MOD to send on the course military officers who are likely to reach the highest ranks. Too often in the past such officers have been deemed to be too busy to attend. The attendance on the same course by both military officers and politicians contributes greatly towards achieving a better mutual understanding between them. Additionally, the scheme for attaching politicians to military units, ships and air stations should also be expanded and carefully targeted.

One of the most important areas of strategy in which greater understanding and agreement is required is that, mentioned earlier, of the appropriate degree of political direction and control of the military. There may still be those in the military who, like US General Franks in Iraq, hanker after an operational level free of any political 'interference'. If so, they need to accept that this apparent military nirvana is actually not as desirable as it seems. It might make military planning and action more pleasurable but it risks such action becoming detached from policy and from the achievement of strategic objectives. It also flies in the face of ministers' right to involve themselves at every level in the making of decisions for which they are ultimately responsible. As Hew Strachan points out, there should be no such thing as a 'politics-free zone'.[96] There is, however, a corollary to this. In order to work well in practice, such political involvement needs to be exercised with inordinately good judgment if it is not to become dysfunctional – for example, in taking-up disproportionate amounts of staff time researching and answering extraneous questions, or inadvertently in constraining initiative and lowering tempo. In some circumstances, for example with great mutual trust between politician and military commander, the lightest of touches will suffice. Defence Secretary Robert Gates describes his relationship with the deployed military commander in Baghdad, General David

[95]Clausewitz, On War, 608.
[96]Hew Strachan, 'The Lost Meaning of Strategy', Survival 37/3 (2005), 33–54, 47.

Petraeus, as a 'partnership', even going as far as to say that 'I would often tell him that Iraq was his battlespace and Washington was mine',[97] although such reassurance did not preclude (nor should it have) detailed oversight and supervision by Pentagon staffs.

The principle of civilian primacy also needs careful interpretation. Good strategy (and good policy) depend on an understanding by all involved that strategy, as a process, involves not only the development and application of ways and means to implement the ends laid down by policy, but also the intellectual activity of balancing ends, ways and means. It requires an acceptance by both politicians and their military advisers that the latter have a duty to engage in the ends/ways/means discourse – what Eliot Cohen describes as 'the unequal dialogue'[98] – including in robust debate, albeit accepting that when formal political direction is given, it should be followed. Where there is a gross imbalance of ends, ways and means, this may require modification of the ends: an adjustment to policy. The danger at one extreme is that, in engaging fully in this debate the military challenge and overstep the line of civilian primacy. At the other extreme is an over-hierarchical interpretation of civilian primacy by the military which leads them to fail to provide sufficiently robust debate, particularly in the face of political pressure. This is a fine line to tread. It is worth noting Thomas E Ricks's observation that '[o]ne of the few predictors of how well a war will go is the quality of discourse between civilian and military leaders'.[99]

Particular care needs to be taken over the selection of those military leaders responsible for providing strategic advice. Although in the UK only one member of the military – the CDS – is formally responsible for providing military advice to the government, the other chiefs of staff play a part in the formulation of that advice and can be directly consulted by the Defence Secretary. Although there are many attributes required by a service chief, competence as an adviser at the grand-strategic level should come at or near the top of the list. Part of such competency is intellect, strategic acumen and wisdom; part is also the moral courage to press unwelcome advice and speak truth to power. In selecting a CDS, politicians are liable to value the former attributes rather more than the latter one. There is, of course, the potential danger that in encouraging senior military officers to be politically savvy, they may be tempted to stray into the political arena, seeking to influence political decisions in an inappropriate way, for example by exerting pressure through the media. The requirement for wisdom, integrity and self-discipline is obvious. A further potential danger may occur when, as in the United States, senior serving or recently retired officers

---

[97]Gates, *Duty*, 49.
[98]Cohen, *Supreme Command*, 242–261.
[99]Ricks, *The Generals*, 214.

are given high-profile political appointments by one political party or leader, thus becoming legitimate targets for public criticism by political opponents.[100] They may, with reason, believe that they have the right of public response. In such circumstances, the issues of civil-military relations raised by Huntingdon and Janowitz may not seem so old-fashioned after all.

In terms of the process of strategy, the current arrangements have rightly attracted criticism for being overly informal, ad hoc and personality-dependant. Lawrence Freedman is not alone in recommending a 'proper, formal process'. As he points out, military advice should not remain in a silo but should be exposed to comment from other government departments much earlier in the process than at present; and this advice should come before a cabinet sub-committee or other group of ministers with, in attendance, 'individuals with experience and authority who can ask good and searching questions'.[101] The military need to know that their advice will be subject to expert scrutiny and challenge, and 'taking military advice should never be made too easy for politicians'.[102] Whether such a process needs to be formally codified and subjected to parliamentary approval, as suggested by de Waal,[103] is less clear.

Finally, a prerequisite for success in the conduct of strategy is recognition at the outset that the political-military dynamic is inherently problematic, that contemporary circumstances make it even more so, and that mutual understanding, respect and trust are at its foundation. All of which is easy to say, but, experience would suggest, much harder to do.

Disclosure statement

No potential conflict of interest was reported by the author.

---

[100]See, for example, Herman Wong, 'Here are the four-star generals Donald Trump has publicly bashed', *Washington Post* (20 October 2017).
[101]Lawrence Freedman, 'On Military Advice', *RUSI Journal* 162/ 3 (2017), 12–19, 18.
[102]Freedman, 'On Military Advice'.
[103]de Waal, *Depending on the Right People*, 35.

Bibliography

Barry, Ben, *Harsh Lessons: Iraq, Afghanistan and the Changing Character of War* (Abingdon: Routledge 2017)

Beadle, Nick, Afghanistan and the Context of Iraq, Michael Clarke, ed., *The Afghan Papers: Committing Britain to War in Helmand 2005-06* (Abingdon: Routledge, 2011)

Bicheno, Hugh, *Razor's Edge: The Unofficial History of the Falklands War* (London: Weidenfeld and Nicholson 2006)

Bobbitt, Philip, *The Shield of Achilles: War, Peace and the Course of History* (London: Penguin 2003)

Bowen, Desmond, The Political-Military Relationship on Operations, Jonathan Bailey, Richard Iron, and Hew Strachan, eds., *British Generals in Blair's Wars* (Farnham: Ashgate 2013), 273–80

Brown, Chris, Multinational Command in Afghanistan – 2006: NATO at the Crossroads, Jonathan Bailey, Richard Iron, and Hew Strachan, eds., *British Generals in Blair's Wars* (Farnham: Ashgate 2013), 217–24

Chilcot, Sir John, *The Report of the Iraq Inquiry: Report of a Committee of Privy Counsellors* (London: Dandy Booksellers Ltd 2016).

Clarke, Michael, The Helmand Decision, Michael Clarke, ed., *The Afghan Papers: Committing Britain to War in Helmand, 2005-06* (London: Royal United Services Institute 2011), 5–29

Cohen, Eliot, *Supreme Command: Soldiers, Statesmen, and Leadership in War* (London: Free Press 2003)

Cowper-Coles, Sherard, *Cables from Kabul: The inside Story of the West`s Afghanistan Campaign* (London: HarperCollins 2011)

Danchev, Alex and Daniel Todman, eds., *War Diaries 1939-1945: Field Marshal Lord Alanbrooke,* (London: Weidenfeld and Nicholson 2001)

Dannatt General Sir, Richard, *Leading from the Front: The Autobiography* (London: Bantam Press 2010)

Dannatt General Sir, Richard, *Boots on the Ground: Britain and Her Army since 1945*: (London: Profile Books 2016)

De Waal, James, *Depending on the Right People: British Political-Military Relations, 2001-10* (London: Chatham House 2013)

Devanny, Joe and Josh Harris, *The National Security Council: National Security at the Centre of Government* (London: Institute for Government 2014)

Elliott, Christopher, *High Command: British Military Leadership in the Iraq and Afghanistan Wars* (London: Hurst & Company 2015)

Fairweather, Jack, *A War of Choice: The British in Iraq 2003-2009* (London: Jonathan Cape 2011)

Fairweather, Jack, *The Good War: The Battle for Afghanistan 2006-2014* (London: Jonathan Cape 2014)

Farrell, Theo, *Unwinnable: Britain's War in Afghanistan 2001-2014* (London: The Bodley Head 2017)

Feaver, Peter, *Armed Servants: Agency, Oversight and Civil-Military Relations* (Cambridge, MA: Harvard University Press 1991)

Feaver, Peter and Richard Kohn, eds., *Soldiers and Civilians: The Civil-Military Gap and American National Security,* (Cambridge, MA: MIT Press 2001)

Finlan, Alastair, War Culture: The Royal Navy and the Falklands Conflict, Stephen Badsey, Rob Havers, and Mark Grove, eds., *The Falklands Conflict Twenty Years On: Lessons for the Future* (London: Frank Cass 2005), 193–212

Franks, Tommy, *American Soldier* (New York: Regan 2004)

Freedman, Lawrence, On Military Advice, *RUSI Journal* 162/3 (2017), 12–19. doi:10.1080/03071847.2017.1345117

Galula, David, *Counterinsurgency: Theory and Practice* (Westport: Prager 2006)

Gates, Robert, *Duty: Memoirs of a Secretary at War* (London: WH Allen 2015)

Gill, Martha, 'Michael Fallon Sets Up Team To Trawl For Chilcot's Buried Recommendations', *Huffington Post* (19 July 2016).

Hastings, Max and Simon Jenkins, *The Battle for the Falklands* (London: Michael Joseph 1983)

Haynes, Deborah, 'US Military Officers Raise Fears over Royal Marine Cuts', *The Times* (25 October 2017).

Healey, Denis, *The Time of My Life: My Autobiography* (London: Penguin 1989)

Heseltine, Michael, *Life in the Jungle: My Autobiography* (London: Hodder and Stoughton 2000)

Huntington, Samuel, *The Soldier and the State: The Theory and Politics of Civil-Military Relations* (Cambridge, MA: Belknap Press of Harvard University 1957)

Jackson, Bill and Dwin Bramall, *The Chiefs: The Story of the United Kingdom Chiefs of Staff* (London: Brasseys 1992)

Jackson, John Hampden, *Clemenceau and the Third Republic* (London: Hodder and Stoughton 1948)

Jackson, Sir Mike, *Soldier: The Autobiography of General Sir Mike Jackson* (London: Corgi 2008)

Janowitz, Morris, *The Professional Soldier: A Social and Political Portrait* (Glencoe, IL: Free Press 1960)

Jeffery, Keith, Sir Henry Hughes Wilson, Baronet, Brian Harrison, Lawrence Goldman and David Cannadine, eds., *Oxford Dictionary of National Biography* (Oxford: Oxford University Press, 2004)

Kiszely, John, Anatomy of a Campaign: The British Fiasco in Norway, 1940 (Cambridge: Cambridge University Press 2017)

Kiszely, John, *Coalition Command in Contemporary Operations* (London: Royal United Services Institute 2008a)

Kiszely, John, Post-Modern Challenges for Modern Warriors, Patrick Cronin, ed., *The Impenetrable Fog of War: Reflections on Modern Warfare and Strategic Surprise* (Westport: Prager 2008b), 129–52

Kiszely, John, The British Army and Thinking about the Operational Level, Jonathan Bailey, Richard Iron, and Hew Strachan, eds., *British Generals in Blair's Wars* (Farnham: Ashgate 2013), 119–30

Krulak, Charles, 'The Strategic Corporal. Leadership in the Three Block War', *Marines Magazine* 83/1(1999).

Ledwidge, Frank, *Losing Small Wars: British Military Failure in Iraq and Afghanistan* (New Haven, CT: Yale University Press 2011)

Maciejewski, Justin, "Best Effort": Operation Sinbad and the Iraq Campaign, Jonathan Bailey, Richard Iron, and Hew Strachan, eds., *British Generals in Blair's Wars* (Farnham: Ashgate 2013), 157–74

Mackay, Andrew and Steve Tatham, *Behavioural Conflict: Why Understanding People and Their Motivations Will Prove Decisive in Future Conflict* (Saffron Walden: Military Studies Press 2011)

Macmillan, Harold, *The Blast of War, 1939-1945* (London: Macmillan 1967)

Massie, Alex, 'You Do the Fighting, I'll Do the Talking', *The Spectator* (20 June 2011).

McColl, John, Modern Campaigning: From a Practitioner's Perspective, Jonathan Bailey, Richard Iron, and Hew Strachan, eds., *British Generals in Blair's Wars* (Farnham: Ashgate 2013), 109–18

McMaster, HR, *Dereliction of Duty: Lyndon Johnson, Robert McNamara, the Joint Chiefs of Staff, and the Lies that Led to Vietnam* (New York: HarperCollins 1997)

Ministry of Defence, *Organising Defence's Contribution to National Strategy*, (London: MOD 2012)

Ministry of Defence, *Joint Doctrine Publication 5-00, Campaign Planning* (London: MOD 2013).

Ministry of Defence, *Making Better Strategy*, (London: MOD 2016a)

Ministry of Defence, *JDP 04. Understanding and Decision Making* (London: MOD 2016b).

Moore, Charles, *Margaret Thatcher: The Authorised Biography*, Vol. *1. Not For Turning.* (London: Penguin 2013)

Oborne, Peter, 'An Officer and a Politician', *The Spectator* (27 May 2000). Available at http://archive.spectator.co.uk/article/27th-may-2000/21/an-officer-and-a-politician.

Perlmutter, Amos, *The Military and Politics in Modern Times: On Professionals, Praetorians, and Revolutionary Soldiers* (New Haven, CT: Yale University Press 1977)

PJHQ, 'Our Organisation and Responsibilities'. Available at https://www.gov.uk/gov ernment/groups/the-permanent-joint-headquarters

Report of the Iraq Inquiry, (London: HMSO 2016). Available at http://www.iraqinquiry. org.uk/the-report/.

Richards, General David, *Taking Command: The Autobiography* (London: Headline Publishing 2014)

Ricks, Thomas, *The Generals: American Military Command from World War II to Today* (New York: Penguin 2012)

Royal College of Defence Studies, *Getting Strategy Right (Enough)*, (London: MOD 2017)

Simpson, Emile, 'America Must Leave the Defeat of Isis to Local States', *Financial Times* (11 September 2014).

Soria, Valentina, Flawed "Comprehensiveness": The Joint Plan for Helmand, Michael Clarke, ed, *The Afghan Papers: Committing Britain to War in Helmand, 2005-06* (London: Royal United Services Institute 2011), 30–48

Strachan, Hew, *The Politics of the British Army* (Oxford: Clarendon Press 1997)

Strachan, Hew, The Civil-Military "Gap" in Britain, *Journal of Strategic Studies* 26/2 (2003), 43–63. doi:10.1080/01402390412331302975

Strachan, Hew, The Lost Meaning of Strategy, *Survival* 37/3 (2005), 33–54. doi:10.1080/00396330500248102

Strachan, Hew, *The Direction of War: A Study in Strategy* (Cambridge: Cambridge University Press 2013a)

Strachan, Hew, Conclusions, Jonathan Bailey, Richard Iron, and Hew Strachan, eds., *British Generals in Blair's Wars* (Farnham: Ashgate 2013b), 327–46

Szayna, Thomas, et aleds. *The Civil-Military Gap in the United States: Does It Exist, Why, and Does It Matter?* (Santa Monica, CA: Rand Corporation 2007)

Tharoor, Shashi, Should UN Peacekeeping Go "Back To Basics"? *Survival* 37/4 (1995), 52–64. doi:10.1080/00396339508442815

UK Parliament, *House of Commons Defence Committee Thirteenth Special Report* (Oct 2007).

Von Clausewitz, Carl, *On War*, edited and translated byMichael Howard and Peter Paret, (Princeton: NJ: Princeton University Press 1976)

Wong, Herman, 'Here are the Four-Star Generals Donald Trump Has Publicly Bashed', *Washington Post* (20 October 2017).

Trigger happy: The foundations of US military interventions

Michael Mayer

ABSTRACT
The United States has repeatedly intervened militarily in situations where tactical success on the battlefield did not translate into meaningful political resolution of the issues triggering the introduction of military force. Although US military interventions are hardly a recent phenomenon, a series of systemic, political and institutional developments over the past several decades have been particularly conducive to the limited use of force as a policy option. These factors have reduced the costs and risks of military intervention, incentivising the use of force in situations when it may not be the optimal policy response.

Among the many poignant moments of the 2017 Ken Burns-Lynn Novick documentary series on the Vietnam War are the audio recordings of two former presidents – John F. Kennedy and Lyndon B. Johnston – acknowledging the likely futility of American military involvement in Vietnam while at the same time escalating the conflict. A similar present-day conundrum – unable to win but unable to exit – was seen in President Donald Trump's announcement in August 2017 that his administration would undertake a renewed effort in the 16-year long war in Afghanistan, committing 3500 additional troops to the limited advisory and counterterrorism operations there. After four US servicemen were killed in counterterror operations in Niger, the *New York Times* published an editorial entitled

> America's Forever Wars' that challenged policymakers to 'take stock of how broadly American forces are already committed to far-flung regions and to begin thinking hard about how much of that investment is necessary, how long it should continue, and whether there is a strategy beyond just killing terrorists.[1]

The academic world has posed similar questions. Considering how a peaceful resolution in Afghanistan might be achieved, Lawrence Freedman argued

---

[1]'America's Forever Wars', *New York Times* (22 October 2017).

that 'the nature of the peace we seek needs to be integrated as a matter of course into any military strategy' and suggested we 'consider what might seriously be achieved through the use of force'.[2]

The United States has a long history of involvement in small wars that stretches back several centuries: policing privateers off the African Barbary Coast during the Jefferson administration, military involvement in the Boxer Rebellion in early twentieth century China, or the numerous military actions and extended occupations in Latin America later that century. As Max Boot chronicles, the US has, since its founding, regularly intervened with armed force, motivated either by geopolitical, economic, humanitarian or punitive reasons.[3] The challenge of connecting political goals and military means is hardly new. The well-known criteria of the 'Weinberger doctrine' for evaluating military interventions, outlined in a speech by Defense Secretary Caspar Weinberger in November 1984, was an attempt to apply the political lessons of the Vietnam War to future decision-making on war and peace. They included viewing the use of force as an option of last resort, as well as having clearly defined political objectives and an understanding of how military force might accomplish those objectives.[4]

In recent decades, however, conditions have become even more conducive for decision-makers to engage in military interventions that clearly deviate from Weinberger's recommendations. The United States remains tactically dominant whenever and wherever force is applied, yet battlefield successes during past decades often have not translated into meaningful political resolution of the issues triggering intervention. Although stability and counterinsurgency operations in Afghanistan and Iraq are perhaps the most glaring and costly examples, recent limited interventions in Libya, Somalia and Yemen also produced questionable long-term strategic benefits despite limited tactical success. Clearly, not all military operations have engendered suboptimal outcomes – the 1991 Gulf War ejecting the Hussein regime from Kuwait and the Balkan conflicts throughout the 1990s were both strategically meaningful and generally successful endeavours. Furthermore, the US was a somewhat reluctant participant in certain multilateral interventions, such as the Balkans or the 2011 Libya operation. Even so, American leaders frequently and often unilaterally employed military force, particularly during the global counterterrorism campaign launched after the terror attacks of 2001.

Interventions lacking political goals achievable through the application of military power are not only expensive mistakes in terms of the nation's

---

[2]Lawrence Freedman, 'Can There be Peace with Honor in Afghanistan?', *Foreign Policy* (26 June 2017).
[3]Max Boot, *The Savage Wars of Peace* (New York: Basic Books 2002).
[4]Eric R. Alterman, 'Thinking Twice: The Weinberger Doctrine and the Lessons of Vietnam', *Fletcher Forum* 10/1 (Winter 1986).

blood and treasure, they also have opportunity costs and second order effects. America's allies in Europe and Asia, while perhaps not necessarily approving of every US intervention, depend on Washington's continued willingness to intervene on their behalf as a crucial component of their security. Particularly in the post-Cold War era, American military power has anchored the international liberal order. But such a powerful tool must be wielded cautiously. The heightened operational tempo of the past several decades strained budgets, personnel, and readiness. Concerns of military 'overstretch' during the final years of the Bush administration gave way to persistent worries of declining readiness, perhaps most aptly illustrated by several serious US Navy ship collisions in 2017 for which an unsustainably high operational tempo resulting in crew exhaustion may have played a major role.[5]

Particularly at a time when the US faces few existential threats, one might expect Washington to take advantage of the luxury its pre-eminence affords it, husbanding its resources and choosing its battles more judiciously. The puzzle here is not about why states go to war, or attempting to distinguish between wars of necessity and wars of choice. Rather, the phenomenon in question is a consistent pattern of military involvement by the dominant power in the internal affairs of other states, often when the use of force is neither geostrategically advantageous nor likely to solve the underlying political issue at stake. Why does the United States repeatedly intervene – or even threaten to intervene – militarily in situations with predictably meagre prospects for successful outcomes and questionable strategic benefits?

This particular question is part of a much broader discussion about the historical global role of the United States. The vast literature relating to this theme ranges from Robert Kagan's *The World America Made* to John A. Thompson's *A Sense of Power*.[6] Importantly, however, this global engagement encompasses a wide range of military, economic and diplomatic activities – not least relating to the creation of international institutions. This article seeks to explore only one aspect of the US as a global actor – the tendency to intervene militarily.

By way of explanation, some have pointed to the evolving nature of armed conflict and a decline in wars that end with a clear-cut winner.[7] Others argue instead that America's use of military force has become detached from any overarching strategy meant to guide it, or warn of an

---

[5]Alex Horton, 'The Navy, Stunned by Two Fatal Collisions, Exhausts some Sailors with 100-hour Workweeks', *Washington Post* (19 September 2017).
[6]Robert Kagan, *The World America Made* (New York: Alfred Knopf, 2012); John A. Thompson, *A Sense of Power: The Roots of America's Global Role* (Ithica: Cornell University Press, 2015).
[7]Robert Mandel, 'Defining Postwar Victory', in Jan Angstrom and Isabelle Duyvesteyn (eds.), *Understanding Victory and Defeat in Contemporary War* (London: Routledge 2007), 13–45.

increasingly militaristic American foreign policy and an overreliance on military instruments in addressing security threats.[8] Given the many small military interventions throughout US history that are contextually unique and vary across a range of geopolitical circumstances and individual officeholders, the causal factors behind these interventionist tendencies can hardly be limited to recent developments, although the rapidly eroding unipolar moment presents a particularly stark contrast between conflicts of strategic necessity and choice.

War and strategy

The decision to employ military force is the most consequential decision political leaders must make. As Robert Art notes, military power is 'the most expensive and dangerous tool of statecraft' and 'the most costly in both treasure and blood'.[9] To be sure, using military force to defend one's own territorial boundaries or protect citizens from attack hardly needs extensive deliberation. When state leaders consider deploying force overseas in a more limited fashion, however, the decision-making process naturally includes an implicit or explicit cost-benefit calculation and a conscious choice either to engage militarily or not. Ideally, these decisions usually entail a broader consideration of the national interests at stake and the threats to those interests that military action will address. This broader analysis of a state's overarching interests, goals and capabilities constitutes a national strategy or 'grand strategy', of which military power is simply one of a number of policy instruments.[10]

Since the passage of the 1986 Goldwater-Nichols Act, US administrations are required to produce and publically release a national security strategy document that attempts to articulate this type of 'whole-of-government' synthesis, although the public nature of the document usually lends itself more to political statement than national strategy. Explicit or not, it is nevertheless within this broader context, taking into account all the available tools of national power, that decisions to use military force should rightly be evaluated. Will military force be more effective than diplomatic or economic instruments, and if all three instruments are used, is the military component precisely calibrated to suit the overarching political goals? When

---

[8]Andrew J. Bacevich, *The New American Militarism* (Oxford: Oxford University Press 2005); Andrew J. Bacevich, *The Limits of Power* (New York: Metropolitan Books 2008); Rosa Brooks, *How Everything became War and the Military Became Everything* (New York: Simon & Schuster 2016); Robert Cassidy and Jacqueline Tame, 'The Wages of War without Strategy', *Strategy Bridge* (5 January 2017); Keith Nightingale, 'Why is America Tactically Terrific but Strategically Shipshod?', *War on the Rocks* (30 September 2015).
[9]Robert Art, *A Grand Strategy for America* (Ithica: Cornell University Press 2003), 4.
[10]Colin Gray, 'Grand Strategy in War and Peace: Toward a broader definition', in Colin Gray (ed.), *Grand strategies in War and Peace* (New Haven: Yale University Press 1991), 5; Sarah Kreps, 'American Grand Strategy after Iraq', *Orbis* 53/4 (2009), 629–645, 633.

can the expected costs and potential consequences of military force be justified, given the entirety of interests and responsibilities of the country?

Stephen Biddle's 2005 assessment of American grand strategy is one example of an externally produced cost-benefit analysis.[11] Biddle evaluated various strategies for addressing two primary sets of threats to the United States: great powers and transnational terrorism. He identified possible desirable end states, the various ways through which the US could seek to achieve them, the costs involved in pursuing these strategies, and the impact on other important national interests such as budgets and diplomatic relationships. Ultimately, Biddle concluded that the 'costs, benefits, and risks cannot be resolved analytically' but depended on 'a series of value judgements'.[12]

Since the end of the Cold War, the judgement of US administrations often favoured military action in situations where the desired end state appeared to be either poorly defined, overly ambitious, or difficult to achieve primarily with military force. This is particularly intriguing because these conflicts seem not to follow the precepts of international relations theories such as realism – which could easily have explained a war of conquest that enhanced US power or its strategic position.[13] Costly interventions without such power-maximising motives are more of a conundrum. Whereas earlier periods in American history might provide a more coherent strategic rationale for using force, US interventions in the post-Cold War era occurred during a period of unprecedented security lacking any overarching incentive to use force in ways American decision-makers wielded US military power. It was in many ways a best-case scenario for *not* engaging in interventionist policies, making the repeated interventions even more interesting.

I argue that even though administrations and security policy elites regularly attempt to formulate overarching strategic concepts in documents that might be useful reference points for decisions to use force, a number of factors at the systemic, state and societal levels enable interventionist policy choices by decision-makers that often undermine these strategic concepts. Systemic factors enabled US policymakers to opt for military intervention without significant resistance from other major powers, and increased concern over non-state threats contributed to an atmosphere of great unpredictability and uncertainty. At the state level, the growing power and influence of the executive branch at the expense of Congress allows for more insulated decision-making processes focused on short-term fixes

---

[11]Stephen Biddle, *American Grand Strategy after 9/11: An Assessment* (Carlisle: Strategic Studies Institute 2005).
[12]Biddle, *American Grand Strategy after 9/11*.
[13]See, for example, John Mearsheimer, *The Tragedy of Great Power Politics* (New York: W.W. Norton 2001) and Stephen Walt, *Taming American Power: The Global Response to American Primacy* (New York: W.W. Norton 2005).

rather than long-term solutions. Additionally, the militarisation of US foreign policy made it more tempting to opt for armed force as a policy tool, increasing the temptation to substitute operational tactics for long-term strategies. Finally, domestic societal factors enabled military adventurism, as a generally willing yet disengaged and polarised public failed to evaluate interventions on their merits.

Systemic complexity

For many analysts, the basic components of US grand strategy have remained fairly consistent over the past century. After pushing westward during the nineteenth century and establishing regional hegemony soon afterwards, the country leveraged its advantageous geographic placement in a peaceful neighbourhood surrounded by weak neighbours and enjoying the stopping power of water. The principal organising strategic goal has been to hinder the rise of a hegemonic power in any key region, but principally on the Eurasia continent. The country's strategic flanks became Europe and Asia-Pacific, which it defended in two world wars and the subsequent Cold War that followed. The collapse of the Soviet Union left the US as the only superpower capable of projecting military power in every region. In the emerging unipolar moment, the US sought to retain its position of primacy through a global network of bases, regional partnerships and alliances combined with overwhelming military capability. Diplomatic efforts were designed to support allies, integrate outliers into the international liberal rules-based order and promote free market democracies, while economic policies generally promoted free trade (but included protectionist policies to placate domestic constituencies) and kept the US and the dollar at the centre of the global economic system.

This overarching strategy can be found implicitly in the sporadically issued national security strategies and quadrennial defense reviews produced by the Pentagon. The approach was perhaps most clearly outlined in the well-known 1992 Defense Planning Guidance document leaked to the *New York Times*.[14] After the Cold War ended, however, the challenge of hindering the emergence of a dominant regional actor in western Eurasia was easier due to Russia's weakened state and China instead became a logical but long-term concern as economic growth fuelled Beijing's rising defense expenditures and growing regional assertiveness.

---

[14]Patrick E. Tyler, 'US Strategy Plan Calls for Insuring No Rivals Develop a One Superpower World', *New York Times* (8 March 1992).

Unipolarity

Unipolarity enabled intractable military interventions because it made the strategic costs appear manageable. Unlike the previous two centuries of small wars, the US had little need to tread carefully and consider global or regional power constellations – the unipolar moment lowered the strategic risks of stumbling into a great power conflict. Lacking any pressing and existential threat, the United States pursued a national strategy based on retaining its dominance and therefore viewed its national interests globally. Its leaders arguably overreacted to non-state and sub-state threats, and fielded a dominant military that could intervene unimpeded in an attempt to resolve those threats.

Unipolarity offered strategic breathing room but the loss of the Soviet Union as an organising framework left decision-makers without clear priorities. During the Clinton administration, Central Intelligence Agency director James Woolsey memorably observed during his confirmation hearing that 'Yes, we have slain a large dragon…But we live now in a jungle filled with a bewildering variety of poisonous snakes. And in many ways, the dragon was easier to keep track of'.[15] Robert Jervis predicted in 1998 that the US would not develop a grand strategy due to the lack of any 'pressing threats' but presciently noted that 'the United States now has less-than-vital interests throughout the world, sufficient power to act on more than a few of them, and an activist ideology'. He worried that 'with less to anchor American policy, smaller events will exert greater influence'.[16]

Each administration in the post-Cold War era has struggled to find an overarching strategic concept. The Clinton administration conveyed the idea of *enlargement*. As National Security Advisor Anthony Lake formulated it in 1993: 'The successor to a doctrine of containment must be a strategy of enlargement, enlargement of the world's free community of market democracies'.[17] The Bush administration, reacting to the shock of 9/11, understandably saw the world through the lens of a global war on terror pursued through military force and intent on – as the administration phrased it – 'building a balance of power that favors freedom'.[18] Facing an increasingly assertive China and mounting domestic budget deficits, President Obama signalled an attempt to correct the perceived 'overstretch' of the Bush years with the release of the 2012 Defense Strategic Guidance entitled 'Sustaining US Global Leadership'.[19] This strategy aimed to limit US commitments overseas by relying more heavily on allies (derisively known

---

[15]Douglas Jehl, 'CIA Nominee Wary of Budget Cuts', *New York Times* (3 February 1993).
[16]Robert Jervis, 'US Grand Strategy: Mission Impossible', *Naval War College Review* LI/3 (Summer 1998).
[17]Thomas Freidman, 'US Vision of Foreign Policy Reversed', *New York Times* (22 September 1993).
[18]George W. Bush, *National Security Strategy of the United States* (Washington DC: White House 2002).
[19]Leon Panetta, *Sustaining Global Leadership: Priorities for the 21st Century* (Arlington: Department of Defense 2012).

as 'leading from behind') as it rebalanced to the Asia-Pacific, strengthened the international liberal order and pursued nation building at home. During his first year in office, President Trump's administration accelerated this trend with a limited multilateral focus, a rhetorical embrace of neo-mercantilist policies and an active military presence in Asia but with limited (primarily counterterrorism) defense commitments elsewhere.

What these strategies largely have in common, however, is a global view of American national interests, broadly defined and motivated by a desire to retain the country's dominant position as the remaining superpower. What they lacked in this pursuit of primacy was, as Nathan Freier noted, 'a more fundamental *ends*-focused, *ways*- and *means*-rationalized, and *risk*-informed grand strategy'.[20] As Freier argued, the US was free in the unipolar moment to 'employ its enormous power reactively and haphazardly', resulting in policy that could be 'more short-sighted, less judicious, more arbitrary, and perhaps more martial than it either should or has to be'.[21] The lack of any peer or even near-peer competitor meant that US military force could – without risking great power conflict – deploy globally against an emerging class of security concerns that posed no direct existential threat to the United States.

Illiberal regional actors with interests at odds with the United States – Iran, Iraq, North Korea, Libya, Syria – were referred to as 'rogue states' representing a direct threat to US national security.[22] The spread of missile and weapons technology combined with the regimes' fiery rhetoric created uncertainty in Washington regarding their ability to deter such states from attacking the US or its allies. Regional stability, which the economic interests of liberal democracies if the free movement of goods were hindered, could be threatened not only by rogue states but also by the collapse of functioning governmental institutions in smaller states. The 2001 Quadrennial Defense Review referred to 'a broad arc of instability' stretching from the Middle East to Northeast Asia, containing 'a volatile mix of rising and declining regional powers', some of which were 'vulnerable to overthrow by radical or extremist internal political forces or movements' or possessed 'the potential to develop or acquire weapons of mass destruction'.[23] The arc of instability, argued the 2004 National Military Strategy, 'serve as breeding grounds for threats to our interests', including asymmetric threats such as terrorism and weapons of mass destruction.[24]

[20]Nathan Freier, 'Primacy Without a Plan', *Parameters* (Autumn 2006), 5–21, 9. Emphasis in original.
[21]Freier, 'Primacy Without a Plan', 12.
[22]Stephen M. Walt, 'Containing Rogues and Renegades: Coalition Strategies and Counter-proliferation' in Victor A. Utgoff (ed.), *The Coming Crisis: Nuclear Proliferation, US Interests, and World Order* (Cambridge, MA: MIT Press 2000), 191–226.
[23]Donald Rumsfeld, *Quadrennial Defense Review Report 2001* (Department of Defense 30 September 2001), 4.
[24]Richard B. Myers, *The National Military Strategy of the United States* (2004), 5.

The emphasis on sub-state threats rather than interstate threats paved the way for a series of counterterrorism operations, security and stabilisation missions, and comprehensive nation-building endeavours. If weak states such as Afghanistan, Somalia or Yemen represented a national security threat due to their attractiveness as a haven for transnational terrorist organisations, government institutions in those states needed to either be restored or created. At the very least, counterterrorism operations required intelligence gathering and operational access for special operations forces and unmanned aircraft. In Iraq, Libya and Syria, US decision-makers deemed autocratic regimes to be a threat to regional security and sought to overthrow them either through direct military intervention (Iraq), an air campaign (Libya) or through limited support for opposition forces on the ground (Syria). In all three cases, the repressive regime served as a check on competing internal forces in complex domestic political situations that erupted as soon as the autocrat was weakened or overthrown.

The application of military force to such situations inevitably led to a more chaotic post-conflict environment, which in turn led to new security threats. It was almost inevitable that armed conflict became more complex under these circumstances, with a desired end state shifting even more profoundly away from decisively defeating an adversary in a conventional interstate conflict and towards more ambiguous metrics for victory. For most conflicts, argues Robert Mandel, the concept of victory can be divided into 'war winning' that resembles the traditional idea of a military victory and 'peace winning' that encompasses the post-conflict reconstruction or transition phase. Historically, 'practitioners have paid much more attention to the first phase than the second, with implicit downplaying of military victory simply being a means to pursue political ends'.[25] For US interventions – particularly after 2001 – the desired end state hinged entirely upon the success of winning the peace, even as planners repeatedly failed to adequately gauge its difficulty and practicality.

The United States easily achieved tactical dominance and amassed battlefield victories throughout its interventions – the war winning phase – but repeatedly struggled to realise the political goals precipitating the use of force. These operations had objectives that could not be conclusively resolved within a specific timeframe, such as denying terrorists safe havens, preventing the spread of weapons of mass destruction, protecting civilian populations or creating democratic institutions. Complicating matters further, political goals often shifted after operations were underway. Furthermore, such objectives were often unrealistic or could not be fully achieved primarily with military force. Ostensibly moving away from nation building in Afghanistan, President Trump declared in August 2017 that 'victory will have a clear definition.

---

[25]Robert Mandel, 'Defining Postwar Victory', 19.

Attacking our enemies, obliterating ISIS, crushing Al Qaeda, preventing the Taliban from taking over Afghanistan and stopping mass terror attacks against America before they emerge'.[26]

Given a definition of victory that relies on tactical goals such as destroying an amorphous terror network but lacks clear political objectives, it is difficult to see how the conflict can ever be successfully and definitively concluded. Alluding to this, US Marine Corps Commandant General Robert Neller reportedly told troops in Helmand province in July 2017: 'I can't guarantee your kids won't be here in 20 years with another old guy standing in front of them'.[27] As Lawrence Freedman noted, 'there is no point in describing an attractive future if there is no obvious way to reach it...opportunities need to be taken to consider what might seriously be achieved through the use of force'.[28] With only China looming as a long-term peer competitor, the United States elevated non-state actors to an existential threat and engaged its military forces accordingly, despite the difficulties of actually resolving those challenges through military power alone.

Strategic uncertainty

The costs and consequences of military interventions are notoriously difficult to predict. While conventional interstate conflicts regularly generate unintended and unforeseen consequences, the pursuit of unclear political objectives that rely on regime change or nation building introduce even higher levels of uncertainty and second order effects. The lack of planning and failure to properly resource the post-conflict phase of the 2003 Iraq War is well known, but the multitude of unknowable variables unleased after intervening in the internal workings of anther state limits the value of such plans.[29] Conducting anything remotely resembling a cost-benefit analysis of an intervention with unpredictable costs and unclear (and perhaps unattainable) benefits is extremely challenging. The shock of the 2001 terrorist attacks reinforced the perception alluded to by Woolsey's testimony that this new unstructured environment increased the risk of strategic surprises. Threat analysis during the Bush administration shifted towards the possible rather than the likely, and influenced decisions ranging from counterterrorism policy to weapons procurement. Principal Deputy Under Secretary of Defense for Policy Ryan Henry in 2005 that 'in order to be able to respond quickly to the unexpected, decision makers will need a broad range of options... rather than focus on specific threats, we focus on capabilities'.[30]

---

[26]'Full transcript and video: Trump's speech on Afghanistan', *New York Times* (21 August 2017).
[27]Thomas Gibbons-Neff, 'It's like Everyone Forgot: On a Familiar battlefield, Marines Prepare for their Next Chapter in the Forever War', *Washington Post* (22 August 2017).
[28]See note 2 above.
[29]See, for example, Thomas Ricks, *Fiasco* (New York: Penguin Press 2006).
[30]Ryan Henry, 'Defence Transformation and the 2005 Quadrennial Defence Review', *Parameters* Winter (2005–2006), 5–15.

On the surface, attempting to hedge against strategic uncertainty by preparing for a broad range of eventualities rather than a specific set of probable threats seems logical. As Jonathan Monten noted, however, 'under conditions of incomplete information, the strategies actors select to achieve their preferences depend on the expectations, or causal beliefs, about the effects of their actions'.[31] This can lead to paradoxical outcomes. Michael Fitzsimmons argued that 'where analysis is silent or inadequate, the personal beliefs of decision-makers fill the void'.[32] This tendency can be self-reinforcing, and decision-makers are often less likely to adjust their thinking when new information emerges, citing the unpredictable nature of the strategic environment. Therefore, the pursuit of flexibility can in fact result in more rigid thinking.

The George W. Bush administration, in judging the sparse intelligence on Iraqi WMD just after the 2001 terror attacks, remained uncertain about its ability to identify and thwart other similar threats before an attack took place. Additionally, leading members of the administration already felt strongly about Saddam Hussein and Congressional legislation from 1999 had made it national policy to support efforts to overthrow the Iraqi autocrat. The lack of clarity regarding the regime's capabilities and intentions inspired two seemingly contradictory tendencies within the administration. On the one hand, President Bush and his inner circle engaged in well-known mechanisms such as worst-case thinking and, as Johannes Rø argued, 'ignore[d] ambiguity and endorse[d] the grimmest interpretation available'.[33]

The United States – leaders and public alike – have often been uncomfortable with ambiguity and sought absolute security. The urge to seek out permanent solutions to intractable problems leaves American decision-makers dissatisfied with unresolved strategic situations, even well-established concepts such as deterrence. Whether the adversary is a regional power seeking nuclear weapons or a small nuclear weapon state in northeast Asia, the US remains uneasy of the prospects of deterring any potential threat to its homeland or its allies with the promise of a devastating counterpunch.[34] Faced with uncertainty, administration officials were sceptical of deterrence, fearful of inaction and not willing to be risk acceptant. On the other hand, the administration engaged in a bit of wishful thinking regarding the operational success of the military intervention and its aftermath, underestimating the resources needed for the stabilisation and reconstruction effort and overestimating the cooperation and political support it would receive from local populations.

---

[31]Jonathan Monten, 'Primacy and Grand Strategic Beliefs in US Unilateralism', *Global Governance* 13 (2007), 130.

[32]Michael Fitzsimmons, 'The Problem of Uncertainty in Strategic Planning', *Survival* 48/4 (2006–2007), 135.

[33]Johannes Rø, *Mechanistic Realism and US Foreign Policy* (London: Routledge 2013), 89.

[34]Colin Dueck, *Reluctant Crusaders* (Princeton: Princeton University Press 2006); Walter Russell Mead, *Special Providence* (New York: Routledge 2002).

Perceptions of strategic uncertainty and a willingness to act on beliefs not supported by evidence contributed to costly military interventions of questionable political benefit for the United States or its allies and partners.

Executive power and militaristic foreign policy

In his sweeping historical analysis of America's rise as a global power, John A. Thompson notes that 'abundant power resources undoubtedly gave the United States the means to exert great influence across the globe. But it did not make it inevitable that it would develop and make use of this capacity'.[35] The contours of the international landscape allow or even incentivise particular strategic decisions, but state leaders' choices ultimately determine policy. Systemic factors enable complex military interventions that may not necessarily serve long-term national interests, but these tendencies have been further augmented in recent decades by state-level factors. Among these are the growing power and influence of the executive branch of the US government and an increasingly institutional and cultural militarisation of US policy that has left the Pentagon a well-funded and omnipresent tool of foreign policy.

Presidential leadership and executive power

Although Congress often acts as a check on the presidency regarding domestic matters, the administration has significant latitude in foreign policy. Congress has generally acquiesced in presidential prerogative on the use of military force. One attempt to wrestle power back to the legislature was the War Powers Resolution of 1973, which requires the President to notify Congress before introducing troops into a conflict and explicit authorisation for longer military engagements. Every president has refused to acknowledge the act's legitimacy, even when seeking congressional authorisation for using force.[36] Presidents customarily procure some type of blanket approval for military action, but have regularly committed forces without any type of formal congressional consent, including Korea (1950–1953), Grenada (1983), Panama (1989), Somalia (1992), the Balkans (1992–1999), Haiti (1993–1996), Libya (2011) and the campaign against ISIS in Iraq (2014).[37]

For counterterrorism operations since 2001, presidents have relied upon congressional legislation authorising the use of force against the perpetrators of the September 11 attacks. The 2001 Authorization for the Use of Military Force (AUMF) gave the president a legal foundation to 'use all

---

[35]Thompson, *A Sense of Power*, p. 230.
[36]Rather than notifying Congress 'pursuant to' the War Powers Resolution, presidents usually employ the phrase 'consistent with'.
[37]James A. Baker et al., *National War Powers Commission Report* (Charlottesville: Miller Center for Public Affairs 2008).

necessary and appropriate force against those nations, organizations, or persons he determines planned, authorized, committed or aided the terrorist attacks...or harbored such organizations or persons, in order to prevent any future acts of international terrorism'.[38] Congress passed a second AUMF in October 2002 specifically authorising the invasion of Iraq, but the 2001 AUMF has served as the legal basis for the broad international counter-terrorism campaign using intelligence operatives, special operations forces and unmanned aerial vehicles. Although the Obama administration argued that the authorisation was sufficient for initiating an air campaign against ISIS in 2014, it nonetheless sent Congress a draft AUMF for its consideration, but the measure never came to a vote. Former senator Jim Webb lamented at the time that

> Congress has become largely irrelevant to the shaping, execution and future of our foreign policy...on the issues of who should decide when and where to use force and for how long, and what our country's long-term relations should consist of in the aftermath, Congress is mostly tolerated and frequently ignored.[39]

Potential constitutional remedies to unauthorised conflicts include impeachment proceedings or failing to authorise spending for military operations, but Congress has consistently failed to avail itself of these measures to halt unpopular conflicts. In an unusual turn of events that resembles a case of the exception proving the rule, President Obama threatened military action in 2012 against the Assad regime in Syria and then sought congressional approval, presenting Congress with a rare opportunity to rebuff the commander-in-chief. Ultimately, the vote never occurred and a diplomatic solution allowed the administration to save face while avoiding the congressional referendum. Obama's decision not to intervene in Syria – in some ways similar to President Bill Clinton's decision not to intervene in Rwanda – became one of the most significant point in his presidency. According to then-Chairman of the Joint Chiefs of Staff General Martin Dempsey, 'our finger was on the trigger...we had everything in place we were just waiting for instructions to proceed' with air strikes.[40] The episode is an outlier in the recent history of US military interventions, but nevertheless illustrates the fact that, ultimately, the decision to use force rests squarely with the president.

The top-heavy structure of the American national security establishment allows for rapid, decisive and flexible responses in the event of a crisis, but this decisiveness comes at a cost. So much power and influence vested in one individual also means a greater risk for mistakes or decisions that run

---

[38]Robert Golan-Vilella, 'A Tale of Two AUMFs', *The National Interest* (27 August 2014).
[39]Jim Webb, 'Congressional Abdication', *The National Interest* (March/April 2013).
[40]Patrice Taddonio, 'The President Blinked: Why Obama Changed Course on the "Red Line" in Syria', *PBS Frontline* (25 May 2015).

counter to the nation's best interests. The small cadre of advisors that comprise a president's inner circle upon which the president seeks advice and political support are subject to the effects of 'group-think' and echo chamber discussions that tend to reinforce existing preconceptions.[41] In many instances, presidents and their advisors decide in favour of interventions and then proceed to make their case to the American public, leveraging its information advantage to control the narrative. An administration may decide to prepare for a military intervention prior to officially making the case before Congress, as was the case with both the 1990 and 2003 Iraq wars. Mobilisation therefore sends a powerful strategic signal to potential adversaries and creates enormous political pressure to support the administration's slow march to war, least the country appear irresolute or individual politicians appear weak on national security issues.

Militarisation of US foreign policy

This coincides with a second important trend in US security policy. The militarisation of US foreign policy is rooted in an institutional resource-based preference for the Department of Defense, but also a deeper cultural and social affinity and faith in the military as an institution. The State Department, for many decades the president's primary advisor on foreign affairs before being overshadowed by the creation of the National Security Advisor role, has struggled institutionally and financially as the Pentagon grew in size and import.[42] Developments during the Trump administration's first year further illustrated this point, with deep cuts to the State Department budget, intentionally unfilled positions, and a management style under former Secretary Tillerson that led to reports of low morale and dysfunction at Foggy Bottom.[43] For national security practitioners, diplomacy is an integral part of the policy toolkit. During the Obama administration, Defense Secretary Robert Gates publically advocated increased funding of the State Department, and Defense Secretary James Mattis once observed wryly during Congressional testimony that the military would need to buy more ammunition if State Department funding were reduced.[44] Even so, the State Department's budget is a fraction of the Pentagon behemoth.

---

[41]For a deeper treatment of the role of elites in executive decision-making on military force, see Elizabeth Saunders, 'War and the Inner Circle: Democratic Elites and the Politics of Using Force', *Security Studies* 24/3 (2015), 466–501.

[42]A.A. Jordan, W.J. Taylor, M.J. Meese, and S.C. Nielsen, *American National Security* (Baltimore: Johns Hopkins University Press 2009), 89–90.

[43]Julie Ioffe, 'The State of Trump's State department', *The Atlantic* (1 March 2017).

[44]Dan Lamothe, 'Retired Generals Cite Past Comments from Mattis while Opposing Trump's Proposed Foreign Aid Cuts', *Washington Post* (27 February 2017).

Defense expenditures have provided the United States with a highly trained and modernised military force, capable of global deployments and able to dominate any potential opponent. This offers policymakers a vast array of military options, including not only conventional land, air, and sea warfighting but also special operations units able to conduct raids against individuals or small groups, drones for surveillance and strike operations, and long-range cruise missiles for deterrence or limited strikes. The temptation of these options is their immediacy – action that can be taken quickly and with visible short-term results – and their global availability, due to a sprawling global network of overseas military installations. Close relationships between large defense contractors, Washington think tanks, military personnel, and Congressional representatives also promote the acquisition and deployment of systems that in turn provide additional military-based options such as drone strikes.

The faith in the utility of military force is also cultural. Andrew Bacevich and Rosa Brooks are among those who have written eloquently about the militarisation of US foreign policy. Bacevich argued in 2005 that 'Americans in our own time have fallen prey to militarism, manifesting itself in a romanticized view of soldiers, a tendency to see military power as the truest measure of national greatness, and outsized expectations regard the efficacy of force'.[45] The American public holds the military as an institution in such high regard that criticism of impending or ongoing operations can often be interpreted as unpatriotic or unsupportive of military personnel. Some scholars argue that political norms emerging during the Cold War, including supporting a strong national security policy and expanded powers for the executive branch, have persisted in the post-Cold War era.[46]

Nearly 80 per cent of Americans in 2016 reported having at least some confidence that the military acts in the public's best interest, with fully one-third of respondents having a 'great deal' of confidence.[47] This faith in the military, while admirable, can encourage interventions when the civilian leadership tasks military leaders with formulating military solutions to perceived threats for which force may not be the optimal policy response. Military force offers a tempting and immediate policy option to complex security challenges while also offering the illusion of 'doing something'. The focus on kinetic operations in counterterrorism interventions appears to favour a similar tactical emphasis without fully integrating these operations into an overarching strategy connected to political outcomes attainable through the application of military force.

---

[45]Bacevich, *The New American Militarism*.
[46]Jane Kellett Cramer 'Militarized Patriotism: Why the U.S. Marketplace of Ideas Failed Before the Iraq War', *Security Studies* 16/3 (2007), 489–524.
[47]Brian Kennedy, 'Most Americans Trust the Military and Scientists to Act in the Public's Interest', *Pew Research* (18 October 2016).

Deference towards military judgements in security policy decision-making is exemplified by the high number of former or current military officers in prominent positions in the Trump White House. Military leaders, ingrained with an optimistic problem-solving 'can-do' attitude, must perform a difficult civil-military balancing act. In his book *Dereliction of Duty*, Lt. General H.R. McMaster criticised the military's complicit role during the escalation of the conflict in Vietnam from 1963 to 1965 due to the failure of its leaders – specifically the Joint Chiefs of Staff – to effectively voice dissenting opinions or concerns.[48] Military leaders voicing dissenting opinions publicly are both noticed and challenged. When the Chairman of the Joint Chiefs of Staff General Martin Dempsey offered a blunt and unusually pessimistic assessment of military intervention in Syria during the Obama administration, it generated debate about civil-military relations and discussions of the balance between providing military advice versus advocating for particular policies.[49]

Supportive, distracted and divided public

Despite the power of the executive and the influence of small groups of elites, broad public support for military interventions remains relevant. As Leslie Gelb wrote in a 1972 article for *Foreign Affairs*, 'For both sides, then, U.S. domestic politics – public support and opposition to the war – was to be the key stress point. American public opinion was the essential domino'.[50] Another of Weinberger's essential criteria for committing troops abroad was the importance of having 'some reasonable assurance we will have the support of the American people and their elected representatives in Congress'.[51]

Military interventions have reflected these political realities, normally avoiding mass mobilisations in favour of more limited deployment of US forces backed by air power and precision guided munitions. This has lessened the human costs of interventions – technology has often become a substitute for blood. The advent of the all-volunteer force and professionalisation of the military means that only a small segment of the US population serves in the military. As the *New York Times* pointed out, 'less than 1 per cent of the population now serves in the military, compared with more than 12 per cent in World War II. Most people simply do not have a family member in harm's way'.[52] Large swaths of the population therefore have little daily contact with military personnel or their families, making the

[48]H.R. McMaster, *Dereliction of Duty* (New York; HarperCollins 1997).
[49]Peter Feaver, 'Civil-military Disagreement? Yes. Crisis? No.', *Foreign Policy* (17 September 2014); Mark Landler and Thom Shanker, 'Pentagon Lays out Options for US Military Effort in Syria', *New York Times* (22 July 2013).
[50]Leslie Gelb, 'The Essential Domino: American Politics and Vietnam', *Foreign Affairs* 50/3 (1972).
[51]Quoted in Alterman, 'Thinking Twice'.
[52]see note 1 above.

human toll of military intervention far less visible. Furthermore, military operations are often financed through deficit spending rather than tax increases, shielding taxpayers from the true cost of military intervention.

Public views on military interventions

Americans are, suggested scholar Richard Sobel, 'ideologically interventionist and operationally anti-interventionist....In principle, people support an aggressive posture, but in practice they are more reluctant to get involved'.[53] An enormous body of academic literature is devoted to understanding the relationship between public opinion and the use of military force. Some studies suggest public support for military action appears to be reliably stronger when the policy objective was to restrain aggression rather than alter the internal politics of another country, while others imply a general cost-benefit analysis weighing the importance of the mission against the number of casualties sustained.[54] Annual surveys by the Chicago Council on Global Affairs find that the public's support for interventions remains consistent with historical patterns – a continued willingness to use force when directly threatened or to solve international problems, particularly if the involvement has a moral component.[55]

In general, an administration has a great deal of latitude to threaten or engage military forces during a crisis. Through its military, intelligence and diplomatic sources, the administration usually possesses far greater quantities of information about any given situation than other branches of government or the general public. Due to this informational monopoly, an administration normally has an initial period of time during which to lay out the arguments for action. The most dependable way to lose support for an ongoing operation, however, is the casualty rate. When military forces begin to suffer casualties, criticism of military action grows and average citizens conduct something approaching a rough cost-benefit analysis. As one study concluded, no other contextual factor appears to have the ability to force a president to end a military operation than its relative value: some casualties can be tolerated for high value missions but only a handful may be acceptable for a low value mission.[56] Jeffrey Record argued that the public 'will support even a costly war

[53]Richard Sobel, *The Impact of Public Opinion on U.S. Foreign Policy Since Vietnam* (New York: Oxford University Press 2001), 16.
[54]Christopher Gelpi, Peter D. Feaver and Jason Reifler, *Paying the Human Costs of War* (Princeton, NJ: Princeton University Press 2006); Bruce Jentleson and Rebecca Britton, 'Still Pretty Prudent: Post-Cold War American Public Opinion on the Use of Military Force', *Journal of Conflict Resolution* 42/4 (1998), 395–417.
[55]Dina Smelz, Ivo Daalder, Karl Freidhoff, and Craig Kafuru, *What Americans think of America first: Results of the 2017 Chicago Council Survey of American Public Opinion and Foreign Policy* (Chicago: Chicago Council 2017).
[56]Louis Klarevas, 'The "Essential Domino" of Military Operations: American Public Opinion and the Use of Force', *International Studies Perspectives* 3 (2002), 417–437.

for a just cause, but they will withdraw their support when they no longer see a reasonable chance for realizing a preferred or acceptable outcome'.[57]

This presents a strategic problem, observed Blechman and Wittes: 'It is rare that it can both make potent threats and retain public support. Potent threats imply greater risks. And the American public's aversion to risk, particularly the risk of suffering casualties, is well known'.[58] When attempting to influence state or non-state actors by issuing threats, American presidents face a dilemma because 'domestic opinion rarely supports forceful policies at the outset, but tentative policies only reinforce the prejudices of foreign leaders and induce them to stand firm'.[59] Since the 1990s, the growing prominence of air power in US warfighting doctrine has allowed presidents to partially escape this dilemma, wielding military force with less risk of casualties. Air campaigns that do not incur US casualties can be sustained without strong opposition, providing a president with a credible threat of force relatively free from public criticism. There are, however, limits to the efficacy of air power for bringing about the political changes on the ground that led to military action in the first place.

Counteracting this aversion to casualties is a deeply ingrained sense of frontier-style justice among Americans. In explaining the nation's strategic choices following the 2001 terror attacks, historian John Lewis Gaddis observed that

> Americans...have generally responded to threats – and particularly to surprise attacks – by taking the offensive, by becoming more conspicuous, by confronting, neutralising, and if possible overwhelming the sources of danger rather than flee from them. Expansion, we have assumed, is the path to security.[60]

Absolute security, even in a globalized world, remains a constant attraction.[61] Rather than wait for enemies to threaten the US homeland, policymakers have preferred to engage those threats overseas. Reinforcing this policy preference was the country's indelible self-image as an exceptional nation founded on universal principles of liberty and morality, with an obligation to spread and defend those values. Wielding military force in pursuit of universal values provided a certain legitimacy to US interventions. As Robert Kagan and William Kristol argued, 'Americans should understand that their support for pre-eminence is as much a boost for international justice as any people is capable of giving'.[62]

---

[57] Jeffrey Record, *Hollow Victory: A Contrary View of the Gulf War* (London: Brasseys 1993), 137.

[58] Barry Blechman and Tamara Cofman Wittes, 'Defining Moment: The Threat and Use of Force in American Foreign Policy', *Political Science Quarterly* 114 (Spring 1999), 27.

[59] Blechman and Wittes, 'Defining Moment', 28.

[60] John Lewis Gaddis, *Surprise, Security and the American Experience* (Cambridge: Harvard University Press 2004), 13.

[61] See also Ulrich K. Preuss, 'The Iraq War: Critical Reflections from "Old Europe"', in Daniel Levy, Max Pensky and John Torpey (eds.), *Old Europe, New Europe, Core Europe* (London: Verso 2005), 169.

[62] Robert Kagan and William Kristol, 'The Present Danger', *National Interest* (Spring 2000), 57–70.

Disengaged and divided public

Despite the global reach of the United States and repeated overseas military interventions, American citizens remain surprisingly uninformed about international affairs and world geography. While hardly a new trend, this lack of knowledge demonstrates widespread civic disengagement on national security issues in ways that have direct policy implications. A 2006 National Geographic poll found that only one-third of young Americans aged 18–24 could identify Iraq on a map of the Middle East despite the ongoing conflict there. In the wake of Russia's 2014 invasion of Ukraine and annexation of Crimea, a national survey queried Americans about their policy preferences regarding Ukraine and then asked to identify the country on world map. Researchers found that only one in six could correctly located Ukraine and, more importantly, the farther their guesses were from the actual location the stronger their support for military intervention.[63] A similar poll conducted three years later, just as the United States and North Korea exchanged heated rhetoric and nuclear threats, found that correctly identifying North Korea on a map corresponded to a greater desire for diplomatic solutions.[64]

In addition to the lack of knowledge about foreign affairs, Americans increasingly view national security issues through a partisan lens. Elite opinion in Washington during the Cold War could usually coalesce around the broad outlines of security policy, even if politics never actually stopped at the water's edge as the conventional wisdom suggested. There has been a growing lack of political unity since 1970s, resulting in the most politically divided electorate in the past century.[65] Not only are the two parties farther apart on issues of national security than was the case several decades ago, they often shift policy stances based on partisanship. For example, presented with a hypothetical question in December 2012 whether they would support military involvement in Syria if the regime there used chemical weapons on its own population, 67 per cent of Republicans answered positively. When the Obama administration considered striking the Syrian regime a year later after its use of chemical weapons, only 22 per cent of Republicans supported US missile strikes. Less than five years later, Republicans overwhelmingly (86 per cent) supported President Trump's missile strikes against Syria in retaliation for another use of chemical weapons against the civilian population.[66] On the

---

[63]Kyle Dropp, Joshua D. Kertzer and Thomas Zeitzoff, 'The Less Americans Know about Ukraine's Location, the More they Want U.S. to Intervene', *Washington Post* (7 April 2014).

[64]Kevin Quealy, 'If Americans Can Find North Korea on a Map, They're More Likely to Prefer Diplomacy', *New York Times* (5 July 2017).

[65]For an excellent overview, see Michael Barber and Nolan McCarty, 'Causes and Consequences of Polarization', in Jane Mansbridge and Cathie Jo Martin (eds.), *Task Force on Negotiating Agreement in Politics* (Washington DC: American Political Science Association 2013).

[66]Aaron Blake, 'Republicans' Transparent, Obama-tinged Flip-flop on Syria', *Washington Post* (11 April 2017).

whole, public views on military interventions are complex, somewhat contradictory, and not always completely rational. A disengaged and less informed
public encourages apathy and acquiescence on the one hand, and opens for
misappropriated heuristics and dangerous polemics to replace reasoned
debate on the other hand.

Conclusions

Although the United States has a long history of military interventionism,
political and institutional developments have made it easier to choose
military force as a foreign policy tool, leading to a persistent pattern of
misapplication of armed force. At the same time, the willingness of US
decision-makers to intervene militarily remains crucial to America's allies in
Europe and Asia. Paradoxically, some of the same structural factors that
enable military interventions can also restrict the use of military force overseas on behalf of alliance commitments, particularly executive dominance of
foreign policymaking and public disengagement. There is little public pressure, for example, to boost US military presence in Europe as part of the
European Reassurance Initiative. In an era of growing budgetary demands
on domestic spending, military budgets cannot grow unbounded. The US
has already discovered that there are limits to its operational capacity even
with a defense budget that dwarfs its nearest competitors. Strategy entails a
husbanding and prioritisation of national resources. Military interventions
conducted without an appreciation of the limits of what military force can
achieve squander these resources and weakens the ability of the United
States to wield force when it really matters.

Disclosure statement

No potential conflict of interest was reported by the author.

Bibliography

Alterman, Eric R., 'Thinking Twice: The Weinberger Doctrine and the Lessons of
    Vietnam', *The Fletcher Forum* 10/1 (Winter 1986), 93–109.
'America's Forever Wars', *New York Times*, 22 Oct. 2017.
Art, Robert, *A Grand Strategy for America* (Ithica: Cornell University Press, 2003).

Bacevich, Andrew J., *The New American Militarism* (Oxford: Oxford University Press, 2005).

Bacevich, Andrew J., *The Limits of Power* (New York: Metropolitan Books, 2008).

Baker, James A., et al. *National War Powers Commission Report* (Charlottesville: Miller Center for Public Affairs, 2008)

Barber, Michael and Nolan McCarty, 'Causes and Consequences of Polarization', Jane Mansbridge and Cathie Jo Martin (eds.), *Task Force on Negotiating Agreement in Politics* (Washington DC: American Political Science Association, 2013), 15–58.

Biddle, Stephen, *American Grand Strategy after 9/11: An Assessment* (Carlisle: Strategic Studies Institute, 2005).

Blake, Aaron, 'Republicans' Transparent, Obama-Tinged Flip-Flop on Syria', *Washington Post*, 11 April 2017.

Blechman, Barry and Tamara Cofman Wittes, 'Defining Moment: The Threat and Use of Force in American Foreign Policy', *Political Science Quarterly* 114 (Spring 1999), 1–30. doi: 10.2307/2657989

Boot, Max, *The Savage Wars of Peace* (New York: Basic Books, 2002).

Brooks, Rosa, *How Everything Became War and the Military Became Everything* (New York: Simon & Schuster, 2016).

Bush, George W., *National Security Strategy of the United States* (Washington DC: White House, 2002).

Cassidy, Robert and Jacqueline Tame, 'The Wages of War without Strategy', *Strategy Bridge*, 5 January 2017.

Cramer, Jane Kellett, 'Militarized Patriotism: Why the U.S. Marketplace of Ideas Failed before the Iraq War', *Security Studies* 16 (2007), 489–524. doi: 10.1080/09636410701547949

Dropp, Kyle, Joshua D. Kertzer, and Thomas Zeitzoff, 'The Less Americans Know about Ukraine's Location, the More They Want U.S. To Intervene', *Washington Post*, 7 April 2014.

Dueck, Colin, *Reluctant Crusaders* (Princeton: Princeton University Press, 2006).

Feaver, Peter, 'Civil-Military Disagreement? Yes. Crisis? No.', *Foreign Policy* 17 September 2014.

Fitzsimmons, Michael, 'The Problem of Uncertainty in Strategic Planning', *Survival* 48/4 (2006–2007), 131–146.

Freedman, Lawrence, 'Can There Be Peace with Honor in Afghanistan?', *Foreign Policy*, 26 June 2017.

Freidman, Thomas, 'US Vision of Foreign Policy Reversed', *New York Times* 22 September 1993).

Freier, Nathan, 'Primacy without a Plan', *Parameters* (Autumn 2006), 5–21.

'Full Transcript and Video: Trump's Speech on Afghanistan', *New York Times*, 21 August 2017.

Gaddis, John Lewis, *Surprise, Security and the American Experience* (Cambridge: Harvard University Press, 2004).

Gelb, Leslie, 'The Essential Domino: American Politics and Vietnam', *Foreign Affairs* 50/3 (1972), 459–475.

Gelpi, Christopher, Peter D. Feaver, and Jason Reifler, *Paying the Human Costs of War* (Princeton, NJ: Princeton University Press, 2006).

Gibbons-Neff, Thomas, 'It's like Everyone Forgot: On a Familiar Battlefield, Marines Prepare for Their Next Chapter in the Forever War', *Washington Post*, 22 August 2017.

Golan-Vilella, Robert, 'A Tale of Two AUMFs', *The National Interest*, 27 August 2014.

Gray, Colin, 'Grand Strategy in War and Peace: Toward a Broader Definition', Colin Gray (ed.), *Grand Strategies in War and Peace* (New Haven: Yale University Press, 1991), 1–11.

Henry, Ryan, 'Defence Transformation and the 2005 Quadrennial Defence Review', *Parameters* Winter (2005–2006), 5–15.

Horton, Alex, 'The Navy, Stunned by Two Fatal Collisions, Exhausts Some Sailors with 100-Hour Workweeks', *Washington Post*, 19 September 2017.

Ioffe, Julie, 'The State of Trump's State Department', *The Atlantic*, 1 March 2017.

Jehl, Douglas, 'CIA Nominee Wary of Budget Cuts', *New York Times*, 3 February 1993.

Jentleson, Bruce and Rebecca Britton, 'Still Pretty Prudent: Post-Cold War American Public Opinion on the Use of Military Force', *Journal of Conflict Resolution* 42/4 (1998), 395–417. doi: 10.1177/0022002798042004001

Jervis, Robert, 'US Grand Strategy: Mission Impossible', *Naval War College Review* LI/3 (Summer 1998), 1–12.

Jordan, A.A., W.J. Taylor, M.J. Meese, and S.C. Nielsen, *American National Security* (Baltimore: Johns Hopkins University Press, 2009).

Kagan, Robert and William Kristol, 'The Present Danger', *National Interest* (Spring 2000), 57–70.

Kennedy, Brian, 'Most Americans Trust the Military and Scientists to Act in the Public's Interest', *Pew Research* 18 October 2016.

Klarevas, Louis, 'The "Essential Domino" of Military Operations: American Public Opinion and the Use of Force', *International Studies Perspectives* 3 (2002), 417–37. doi: 10.1111/insp.2002.3.issue-4

Kreps, Sarah, 'American Grand Strategy after Iraq', *Orbis* 53/4 (2009), 629–45. doi: 10.1016/j.orbis.2009.07.004

Lamothe, Dan, 'Retired Generals Cite past Comments from Mattis while Opposing Trump's Proposed Foreign Aid Cuts', *Washington Post* 27 February 2017.

Landler, Mark and Thom Shanker, 'Pentagon Lays Out Options for US Military Effort in Syria', *New York Times* 22 July 2013.

Mandel, Robert, 'Defining Postwar Victory', Jan Angstrom and Isabelle Duyvesteyn (eds.), *Understanding Victory and Defeat in Contemporary War* (London: Routledge, 2007), 13–45.

McMaster, H.R., *Dereliction of Duty* (New York: HarperCollins, 1997).

Mead, Walter, *Russell, Special Providence* (New York: Routledge, 2002).

Mearsheimer, John, *The Tragedy of Great Power Politics* (New York: W.W. Norton, 2001).

Monten, Jonathan, 'Primacy and Grand Strategic Beliefs in US Unilateralism', *Global Governance* 13 (2007).

Myers, Richard B., *The National Military Strategy of the United States* (2004).

Nightingale, Keith, 'Why Is America Tactically Terrific but Strategically Shipshod?', *War on the Rocks* 30 September 2015.

Panetta, Leon, *Sustaining Global Leadership: Priorities for the 21st Century* (Arlington: Department of Defense, 2012).

Preuss, Ulrich K., 'The Iraq War: Critical Reflections from "Old Europe"', Daniel Levy, Max Pensky, and John Torpey (eds.), *Old Europe, New Europe, Core Europe* (London: Verso, 2005), 50-72.

Quealy, Kevin, 'If Americans Can Find North Korea on a Map, They're More Likely to Prefer Diplomacy', *New York Times* 5 July 2017.

Record, Jeffrey, *Hollow Victory: A Contrary View of the Gulf War* (London: Brasseys, 1993).

Ricks, Thomas, *Fiasco* (New York: Penguin Press, 2006).

Rø, Johannes, *Mechanistic Realism and US Foreign Policy* (London: Routledge, 2013).

Rumsfeld, Donald, *Quadrennial Defense Review Report 2001* Department of Defense 30 September 2001.

Saunders, Elizabeth, 'War and the Inner Circle: Democratic Elites and the Politics of Using Force', *Security Studies* 24/3 (2015), 466–501. doi: 10.1080/09636412.2015.1070618

Smelz, Dina, Ivo Daalder, Karl Freidhoff, and Craig Kafuru, *What Americans Think of America First: Results of the 2017 Chicago Council Survey of American Public Opinion and Foreign Policy* (Chicago: Chicago Council, 2017).

Sobel, Richard, *The Impact of Public Opinion on U.S. Foreign Policy since Vietnam* (New York: Oxford University Press, 2001).

Taddonio, Patrice, 'The President Blinked: Why Obama Changed Course on the "Red Line" in Syria', *PBS Frontline* 25 May 2015.

Tyler, Patrick E., 'US Strategy Plan Calls for Insuring No Rivals Develop a One Superpower World', *New York Times* 8 March 1992.

Walt, Stephen, *Taming American Power: The Global Response to American Primacy* (New York: W.W. Norton, 2005).

Walt, Stephen M., 'Containing Rogues and Renegades: Coalition Strategies and Counter-Proliferation', Victor A. Utgoff (ed.), *The Coming Crisis: Nuclear Proliferation, US Interests, and World Order* (Cambridge, MA: MIT Press, 2000), 191–226.

Webb, Jim, 'Congressional Abdication', *The National Interest* March/April 2013.

Weak party escalation: An underestimated strategy for small states?

Jan Angstrom and Magnus Petersson

ABSTRACT

In this article, we develop the strategic rationale behind weak party escalation against stronger adversaries. There are, we suggest, four main strategies: to *provoke* a desired over-reaction from the stronger adversary; to *compartmentalize* conflict within a domain in which the weak party has advantages; to *carve a niche* with a stronger ally, and to forge a *reputation* of not yielding lightly. Spelling out these different logics contributes to the literature on small state strategies and escalation. It also suggests, contrary to much of the existing literature, that it can be rational for weak parties to escalate against great powers.

Why do weak parties escalate against stronger adversaries? The question is highly topical given the sabre rattling of North Korean leader Kim Jong-un against the US in 2017. Weak-parties', especially small states', escalation against stronger powers is commonly understood to be irrational behaviour emanating from misperceptions, mistakes or downright madness. We suggest, by contrast, that weak party escalation against stronger adversaries can be strategic, and thus rational. Still, even if we argue that weak party escalation in asymmetric relations can be a strategic choice, it does not mean that we suggest that it is empirically prevalent, nor that the outcome is beneficial for the weak. Instead, we suggest that weak party escalation only is rational when specific conditions hold and once those conditions do not hold, the weak party should not escalate.

Although weak party escalation is not as widespread as strong-party escalation, the weak occasionally escalate. For example, in 1708, during the Great Northern War (1700–1721), the Swedish King Charles XII invaded a much stronger Russia to preserve Sweden's great power status in the Baltic Region. The invasion was successful to begin with, but eventually became a disaster for

Sweden. The Swedish Army was destroyed in the battlefields of Poltava, and the era of Sweden as a great power ended.[1] Furthermore, in 1862, Confederate General Stonewall Jackson's Shenandoah Valley campaign effectively shifted the entire US Civil War front away from Confederate capital Richmond, and briefly, threatened to open a route towards Washington for the Confederate army. Moreover, in the so-called war of continuation 1941–1944 during the Second World War, Finnish forces attacked the Soviet Union to reclaim Karelia – a part of Eastern Finland that had been lost in the Winter War a year earlier. This seems counter-intuitive given that the Soviet Army at the time numbered 3,5 million soldiers, at least 7-times the size of the Finnish army.

Although there are many studies of escalation, in particular, those that theorize nuclear war, there is a surprising lack of knowledge of interstate weak party escalation. At large, the literature on escalation is divided into two debates. First, there seems to be profound uncertainty about how to understand escalation. On the one hand, in much peace and conflict research, escalation is conflated with causes of war.[2] This understanding is closely linked to the notion of the security dilemma,[3] and wars erupting even in situations where the actors do not prefer war. Escalation, in this understanding, happens regardless if the actors desire it. The metaphor of a self-propelling escalator underpins this notion,[4] and it has led theorists to assume that war could spiral out of control. Crises escalating into all-out war was seen as inevitable and a worst-possible outcome. Escalation is very much understood as a noun according to these scholars.

On the other hand, within strategic studies, scholars tend to understand escalation as *deliberate* changes in quantitative or qualitative dimensions of the use of force.[5] By approaching escalation as intentional, it also becomes clear that escalation is understood as a verb in this tradition. As such, escalation is not understood to be only about the 'first shot' in a war, but rather how the actors interact and how the dynamics of violence unfold. The central puzzle thus involves how the variation of the level of violence in time and space influences the degree of success in battles, campaigns and wars. As will be evident below, we follow this second understanding, i.e., escalation is not limited to the first act of violence, but neither is it limited to what goes on within war either. The point being that the actor, when faced with a decision between escalation or de-escalation, does not know whether or not the outcome will be war (or continuation of war) since, at the time of the decision, it is impossible to know the

[1] Ragnhild Marie Hatton, *Charles XII of Sweden* (London: Weidenfield and Nicolson 1968).
[2] T.V. Paul, *Asymmetric Conflicts: War Initiation by Weaker Powers* (Cambridge: Cambridge University Press 1994); Geoffrey Blainey, *The Causes of War*, 3rd edn. (New York: The Free Press 1988).
[3] John Herz, 'Idealist Internationalism and the Security Dilemma', *World Politics* 2/2 (1950) 171–201.
[4] Lawrence Freedman, 'On the Tiger's Back: The Development of the Concept of Escalation', in Roman Kolkowicz (ed.), *The Logic of Nuclear Terror* (London: Allen & Unwin 1987), 109–52.
[5] Thomas Schelling, *Arms and Influence* (New Haven, CT: Yale University Press 1966).

response of the adversary. Escalation can, therefore, go on within war, it can ignite a war, and it can prevent war.

Second, within strategic studies, there are further debates related to the dynamics of war and how this is related to escalation. In particular, these debates have centred around what kind of military means are related to escalation and why some wars escalate more than others. Here, there is a huge literature on the dangers of escalation from limited war into nuclear war,[6] but also on low-level escalation in counter-insurgencies and civil wars.[7] Most of what we know of theories of escalation is related to Schelling and his work on symmetric power relations. When it comes to asymmetric relations, research is scarce with the possible exception of intrastate wars.[8] Below, when elaborating on how rationality and escalatory behaviour is related and conceptualized, we also discuss these debates more carefully.

The contribution of this article is to analyze and develop different logics of weak party escalation in interstate asymmetric relations. This is a distinct contribution to the literature on escalation, but it is also relevant for the literature on small state strategy more broadly (since small states often find themselves on the weak end of asymmetric relations). In our analysis, we draw upon multiple cases to illustrate and develop different logics of weak party escalation. A stronger focus on a single case would not allow us to identify variation in rationales of weak party escalation. Intentionally, and following how we conceptualize escalation, these cases also include a variation of escalatory processes from sabre rattling verbal behaviour to all-out military action. Why and how do weak parties escalate in their conflicts with more powerful adversaries? How does escalation work in asymmetrical relations? Do the weak always have to suffer what they must – as Thucydides famously put it – or are there cases where the weak can create escalation dominance in a narrow subset of the 'escalatory ladder' (in Kahn's words) or cases where weak party power escalation can provoke or create great power overstretch? [9]

The article continues as follows. First, we will outline the strategic logic of four rationalist explanations of weak party escalation. In short, weak parties escalate, first, to *provoke* a desired over-reaction from the stronger

---

[6]Herman Kahn, *On Thermonuclear War* (Princeton, NJ: Princeton University Press 1960); Bernard Brodie, *Escalation and the Nuclear Option* (Princeton, NJ: Princeton University Press 1966); Fred Kaplan, *The Wizards of Armageddon* (Stanford, CA: Stanford University Press 1983).

[7]David Lake and Donald Rotchild (eds.), *The International Spread of Ethnic Conflict: Fear, Diffusion, and Escalation* (Princeton, NJ: Princeton University Press 1998); Isabelle Duyvesteyn, 'The Escalation and De-escalation of Irregular War: Setting Out the Problem', *Journal of Strategic Studies* 35/5 (2012) 601–12.

[8]Exceptions regarding terrorism and anti-terrorism include Duyvesteyn, 'The Escalation and De-escalation of Irregular War', 601–12, and regarding small state nuclear powers versus strong nuclear powers, Carmel Davis, 'An Introduction to Nuclear Strategy and Small Nuclear Powers: Using North Korea as a Case', *Defence Studies* 9/1 (2009) 93–117.

[9]Herman Kahn, *On Escalation: Metaphors and Scenarios* (New York: Praeger 1965).

adversary. This over-reaction, in turn, can be beneficial for the weak party as it may trigger outside help. Second, weak parties escalate if they can *compartmentalize* conflict within a domain in which they can, despite their overall inferiority, maintain escalatory dominance. Third, weak parties escalate by creating a *division of labour* with a stronger ally. Fourth, weak parties escalate to forge a *reputation* of not yielding lightly. This final logic suggests that even if the immediate consequences of escalation are negative, the long-run benefits of maintaining a reputation of being steadfast can be more important.

After outlining these logics, we will elaborate on the dilemmas of the weak when escalating. In short, there are dangers associated with being stood up as a result of being perceived as the aggressor, with limiting, thus controlling war, and with short-term costs being difficult to assess correctly, potentially leading to a situation where the long-term benefits of escalation may not be reaped. In this section, we further stress that we understand escalation as intended, rational, strategic action, but with uncertain outcomes. In short, just because an action is intended and indeed rational, does not guarantee a happy ending. Weak party escalation is always fraught with dangers and these dangers also make weak party escalation empirically rare.

Finally, in the concluding part of the article, we will argue that it is important for policymakers of the weak to realize that escalation can be a rational choice in a conflict, increasing the strategic options for such policymakers to achieve their strategic goals.

Rationalism, strategy and escalation

Within rationalist theories of war, escalation is commonly understood to be a key operating dynamic in militarized crises and wars. Strategy at large, and escalation, in particular, can of course also be understood from non-rationalist theories. Fisherkeller, for example, argues that weak party escalation is the result of misperceptions or a particular form of especially belligerent strategic culture.[10] In this article, however, we start off by anchoring strategy within rationalism.

The common way to understand strategy makes it abundantly clear that the concept relies heavily upon rationalist assumptions. Conceptualizing strategy as striking a balance of *ends, means,* and *ways* in war, for example, implies that military means are used for political ends and thus an instrumental understanding. The ways in which military means are used, moreover, also relies on a rationalist notion of using means effectively, rather than randomly. Strategic analysis, therefore, often becomes an effort of making

---

[10]Michael P. Fischerkeller, 'David versus Goliath: Cultural Judgements in Asymmetric Wars', *Security Studies* 7/4 (1998), 1–43.

decisions understandable by referring to the political ends of the war. Naturally, this creates important debates about how to conceptualize political ends or national interests. Whereas scholars such as Kenneth Waltz argues that states are interested in status quo, others such as Hans Morgenthau and John Mearsheimer suggest that states seek to maximize power.[11]

Strategists commonly emphasize that strategy in practice is about creating power and achieving maximum political leverage despite possessing scarce resources. This further strengthens the ties to rationalism. It is precisely because strategy is conducted under such conditions that it is important that appropriate military means are created and that the ways they are used are efficient. The third connection between strategy and rationalism derives from the fact that strategy is never conducted in isolation. In any dynamic, interactive situation, actors on both sides have interests to maintain some secrets and neither actor can be sure of each other's intentions. Escalation, in the sense of attempting to destroy the opponent's military means, is attractive since it reduces uncertainty in the decision-making.[12]

For rationalists, for example, bargaining theories of war,[13] war can be fruitfully approached as patterns of escalation and de-escalation and the commitments that are communicated in such patterns. In the extreme, another generation being sent 'over the top' in the trenches during the battles of Somme or Verdun were costly signals to the opponent of how credible were the opponents' desire for victory. Hence, to understand the dynamics of war, we must focus on escalation and de-escalation. Actors with so-called 'escalation dominance', according to Kaplan and Schelling, will ultimately have a huge advantage in an upcoming political and military clash since the ability to escalate more than your opponent in various stages of the war will increase your chances of dominating the outcome.[14]

These ideas, originating in Clausewitz' thinking, suggest that escalation is inherent in war since in every war there is a dynamic interaction in which each actor's behaviour is interdependent on the opponent's actions. This gradual escalation of violence, Clausewitz posited, was unnecessary in

---

[11]Kenneth Waltz, *Theory of International Politics* (New York: McGraw-Hill 1979); Hans Morgenthau, *Politics Among Nations: The Struggle for Power and Peace* (New York: McGraw-Hill 1948); John Mearsheimer, *The Tragedy of Great Power Politics* (New York: Norton 2001).

[12]Richard K. Betts, 'Should Strategic Studies Survive?', *World Politics* 50/1 (1997), 12; James Gow, *The Serbian Project and its Adversaries: A Strategy of War Crimes* (London: Hurst 2003), 16–17; Benedict Wilkinson and James Gow (eds.), *The Art of Creating Power: Freedman on Strategy* (London: Hurst, 2017).

[13]R. Harrison Wagner, 'Bargaining and War', *American Journal of Political Science* 44/3 (2000), 469–84; Robert Powell, 'Bargaining Theory and International Conflict', *Annual Review of Political Science* 5 (2002), 1–30; Dan Reiter, 'Exploring the Bargaining Model of War', *Perspectives on Politics* 1/1 (2003), 27–43; Branislav Slantchev, *Military Threats: The Costs of Coercion and Price for Peace* (Cambridge: Cambridge University Press 2011); Mark Kilgour and Frank Zagare, 'Explaining Limited Conflicts', *Conflict Management and Peace Science* 24/1 (2007), 65–82.

[14]Paul, *The Strategy of Limited Retaliation* (Princeton, NJ: Woodrow Wilson School 1959); Schelling, *Arms and Influence*.

practice, instead suggesting that it was rational for the actors to use over-whelming force immediately. In a much-quoted passage, Clausewitz sug-gested that 'there is no logical limit to the application of that force. Each side, therefore, compels its opponent to follow suit; a reciprocal action is started which must lead, in theory, to extremes'.[15]

It was this idea of immediate maximum escalation that spawned the idea that whomever had escalation dominance, i.e., that could muster the most military power, was already a winner. And this assumption led to the conclusion that it was irrational for actors not wielding escalation domi-nance to escalate. Hence, in general, weak parties ought not to escalate. Clausewitz, however, also raised the problems of over-escalating by intro-ducing the concept of the culminating point, i.e., that there is a point where even the nominally stronger party over-stretches its resources and becomes vulnerable to counter-attack.[16]

The causal mechanism in much research on escalation, therefore, relates to the threat of future damage that the stronger side can inflict upon the weaker. As such, escalation as a military strategic decision is more related to coercion and deterrence than brute force. If you are able to escalate more than your opponent, you will be in a position to hurt your opponent and be the last one standing when the dust settles. Weaker parties do not, to use Rupert Smith's words, possess a natural, numerical ability 'to escalate to success'.[17] But occasionally they do it anyway. So why do weak parties escalate?

Theorizing weak-party escalation in asymmetrical relations

In this section, we develop the strategic logic of four rationalist explanations of weak party escalation. Before outlining these logics, however, it is neces-sary to elaborate more carefully on the concept of escalation and asymme-try. We deviate slightly from Smoke's definition of escalation and suggest that one ought to include actions 'that crosses saliencies' both within and outside war. Smoke's influential definition understood escalation as an 'action that crosses a saliency which defines the current limits of a war, and that occurs in a context in which the actor cannot know the full consequences of his action, including particularly how his action and the opponent's potential reaction(s) may interact to generate a situation likely to induce new actions that will cross still more saliences'.[18]

[15]Carl von Clausewitz, *On War*, translated by Michael Howard and Peter Paret (Princeton, NJ: Princeton University Press 1976), 77. Mike Smith, 'Escalation in Irregular War: Using Strategic Theory to Examine from First Principles', *Journal of Strategic Studies* 35/5 (2012), 613–38.

[16]von Clausewitz, *On War*, 638–9.

[17]Rupert Smith, *The Utility of Force: The Art of War in the Modern World* (London: Allen Lane 2005).

[18]Richard Smoke, *War: Controlling Escalation* (Cambridge, MA: Harvard University Press 1977), 35.

Escalation, in other words, is about increasing the level of, changing the type of or altering the geographical confinements of violence. This boundary crossing behaviour can go on both within and outside war. Typically, it, therefore, involves actions such as opening up a new front or increasing the number of troops engaged in operations but also breaking other limits such as starting to bomb previously not attacked cities. The main reason for our inclusion of escalatory behaviour outside war is that at the time of the decision, actors do not know whether or not an action will ignite war or, for that matter, end war. However implausible it may sound, we suggest that escalation can be a rational strategic choice for weaker parties.

We define asymmetrical relations in terms of nominal military power, i.e., the size of a state's military organization.[19] We follow Arreguin-Toft's ratio of 1:5 to identify strong-weak dyads.[20] A ratio of 1:5 and the weak still escalates is a 'hard' test of the causal logic. In many Western military doctrines and among military theorists, in more general, a common rule of thumb is that you can go on the offence only if you possess an advantage of 3:1.[21] Being at a disadvantage of 1:5, therefore, ought not to generate escalation.

As the above discussion of the role of escalation in strategy showed, two conditions stand out when explaining escalation in general. In our theorizing on weak party escalation, we draw upon the same two conditions. First, the degree to which the weaker party can rely upon external support will heavily influence the rationale of escalation. Second, the degree to which the weaker party can be said to still maintain escalation dominance will also heavily influence the rationale of escalation. Figure 1 (below) outlines the strategic logic.

In the figure, we can also identify why weak party escalation is quite rare. Few small, exposed states have powerful allies that are able to back them up should the small state escalate in a too rash manner. Therefore, strategic weak party escalation in general, might work better in an alliance context. After outlining the logic of escalation, we return to a discussion on why weak party escalation is empirically rare, despite it sometimes can be rational and strategic. The simple two-variable theory thus can explain *variation* in weak party escalation.

---

[19]Paul, *Asymmetric Conflicts*; Ivan Arreguin-Toft, *How the Weak Win Wars: A Theory of Asymmetric Conflict* (Cambridge: Cambridge University Press 2005); Jan Angstrom, 'Evaluating Rivalling Interpretations of Asymmetric War and Warfare', in Karl-Erik Haug and Ole Jorgen Maao (eds.), *Conceptualising Modern War* (New York: Columbia University Press 2011); Andrew Mack, 'Why Big Nations Lose Small Wars: The Politics of Asymmetric Conflict', *World Politics* 27/2 (1975), 175–200.
[20]Ivan Arreguin-Toft, 'How the Weak Win Wars: A Theory of Asymmetric Conflict', *International Security*, 26/1 (2001), 96.
[21]J.F.C. Fuller, *The Foundaitions of the Science of War* (London: Hutchinson 1926).

External support

	Yes	No
Yes	Division of labor	Compartmenta-lization
No	Provocation	Reputation

Escalation dominance (label at left, between Yes and No rows)

Figure 1. The strategic logic of weak party escalation against stronger adversaries.

The strategic logic of alliances – creating a division of labour

Weaker parties that can rely upon alliances or at least friends that together possess escalation dominance have incentives to escalate. This is a well-recognized logic in which the weaker party in an alliance can be more daring than it would have been otherwise, provided that it can trust allies to come to its rescue if the opponent reacts in an unforgiving manner.

The logic is aptly captured by the Finnish decision to attack a Soviet Union the was at least 7 time stronger than them in the War of Continuation. At the time, Finland – although not formally in an alliance with the Axis Powers – could use Nazi Germany's attack on the Soviet Union in June 1941 to launch its own offensive a few weeks later to reclaim territory it had lost during the Winter War.[22] Rather than being in a formal alliance, one can certainly claim that Finland's military aims were aligned with the Axis Powers' and that it was more daring than it otherwise would have been. As such, creating a division of labour and escalating against a stronger adversary while it is pre-occupied in armed conflict with an equal resembles so-called bandwagoning, where weak states rally with stronger states to earn short-term gains.[23]

Despite being nominally weaker it is rational for the weaker, party to escalate since it relies on external support that possesses escalation dominance, if – and only if – it can rely upon the stronger allied partner to partake in the war. It is also important to recognize that escalation may come in different forms. All-out offensive actions against a stronger party may be unusual, but as the case of Finland in 1941 – and, for example, Prussia against the 5-times stronger alliance of Austria, France and Russia in the Seven Year's War 1756–1763 – prove they still occur.[24]

---

[22]Carl Henrik Meinander, 'Finland', in David Stahel (ed), *Joining Hitler's Crusade: European nations and the invasion of the Soviet Union, 1941* (Cambridge: Cambridge University Press, 2017).
[23]Randall L. Schweller, 'Bandwagoning For Profit: Bringing the Revisionist State Back In', *International Security* 19/1 (1994), 72–107.
[24]Franz A.J. Szabo, *The Seven-Years War in Europe, 1756–1763* (London: Routledge 2008).

More commonly, escalation may take other forms. We can see this logic in Iceland's decision to escalate conflict against the United Kingdom during the so-called Cod Wars in the late 1950s to mid-1970s. Although the Cod Wars did not really play out as high-intensity conventional wars, they still represent interesting cases for our purposes, since they involve asymmetrical relations and weak party escalation. Not only did Iceland further its fishing rights successfully throughout the Cod Wars despite being vastly outnumbered, they were also continuously escalating the wars, while throughout using subtle – or not so subtle – suggestions that they would leave NATO.[25]

Had Iceland left NATO, of course, the whole NATO Northern Flank would have become much harder to defend. This was down to not only threats about buying Soviet frigates (instead of American) or closing the Keflavik Airbase, but the integrity of the so-called GIUK Gap was in danger. The GIUK Gap, of course, was of strategic importance for the United States and its ability to reinforce NATO forces in Europe since it was a critical choke point to prevent Soviet nuclear submarines free access to the Atlantic.[26] Through successfully deterring the conflict to escalate into other domains or from increasing in intensity, the Icelandic government despite its inferiority was able to escalate more than it otherwise would have.

The strategic logic of compartmentalization

It can also be rational and strategic for weaker parties to escalate against nominally stronger adversaries – in cases where they lack external support. In fact, if the weaker party can isolate the conflict into a domain where it has escalation dominance, it effectively reverses the overall power structure. In theory, therefore, the weaker party becomes the stronger party as long as it can maintain and limit escalation to within the domain in which it has escalation dominance. To pursue such a strategy requires that the punishment for adversary cross-domain escalation must be higher than the cost of losing the limited contest. In other words, compartmentalization requires the ability to deter opponent escalation in other domains.[27]

To some extent, we can think of Britain's escalation against Napoleonic France – primarily through the Royal Navy – as an example of this logic.

[25]Jacob Børresen, *Torskekrig!: Om forutsetninger og rammer for kyststatens bruk av makt* (Oslo: Abstrakt 2011).

[26]Rolf Tamnes, *The United States and the Cold War in the High North* (Aldershot: Palgrave 1991).

[27]The literature on deterrence is vast. We follow a standard definition of deterrence as 'the persuasion of one's opponent that the costs and/or risks of a given course of action he might take outweigh its benefits'. Alexander George and Richard Smoke, *Deterrence in American Foreign Policy* (New York: Columbia University Press, 1974), p. 11.

Once Britain and the remaining part of the Second Coalition had been defeated on land by the superior French forces, the Royal Navy intensified its attacks on French ships at sea. On the High Seas, the Royal Navy had escalation dominance and as long as the war was fought in that domain, Britain could escalate at will despite its huge overall inferiority. The strategy had its ups and downs. To prevent France from invading (a cross-domain escalation that Britain would not have survived), Britain had to pursue a vigorous diplomatic campaign to, on seven occasions, form coalitions that threatened the French land supremacy. Only the last two were eventually victorious against Napoleon.[28]

Hence, the strategy of escalation against the nominally stronger France was fraught with dangers for the British. Despite being nominally weaker, though, Britain relied on its navy to escalate the war. Keeping the war within the domains of the High Seas meant that Britain could fight the war on favourable terms.

The logic of compartmentalization relies upon a calculus about being able to deter cross-domain escalation on behalf of the adversary, while escalating in one's preferred domain. This idea has long pedigree; Brodie held that 'controlling escalation is an exercise in deterrence'.[29] At the height of the Cold War nuclear balance, much effort was made to establish what one hoped to be 'firebreaks' that could check and prevent limited war escalating into all-out nuclear war. The same logic applies for the weak party in an asymmetrical escalation phase. As we could see in the case above, Britain was able to deter a French land invasion in which it would have been inferior, while maintaining and increasing the war effort on the High Seas.

A more current example of potential compartmentalization is Russia's ability to escalate in the nuclear domain, to avoid losing conventionally, because of NATO's conventional superiority. Among experts in the West, Russia is believed to use nuclear weapons in a conflict according to the concept of 'escalate-to-deescalate'. That means that if Russia faces NATO's conventionally stronger forces, they could turn to limited use of nuclear weapons, to convince NATO that further nuclear escalation would proceed if NATO choose not to back down.[30]

Another potential future (or maybe contemporary) case of compartmentalization escalation can be an escalation in the cyber-domain. The advantage a weak party can have in cyber-escalation is that the weak party can have actual dominance in the cyber-domain and – for example – neutralize or destroy the stronger party's vital resources (such as communications

---

[28]Roger Knight, *Britain against Napoleon: The organization of Victory* (London: Penguin, 2014).
[29]Brodie quoted in Freedman, 'On the Tiger's Back', 125.
[30]Ulrich Kühn, *Preventing Escalation in the Baltics: A NATO Playbook* (Washington, DC: Carnegie, 2018).

systems) and/or deniability and claim that a cyber-attack was not sanctioned by the state. Since accountability is difficult in the cyber-domain, a weak party can escalate with less fear of retaliation and less fear of an opponent's escalation in other domains.[31]

The strategic logic of provocation

The third strategic rationale of weak party escalation builds upon conditions where the weak party does not have escalation dominance, but can, to some extent, rely upon external support. The idea here is not, as in carving a niche, that the weaker party takes part in an overall opportunistic, predatory escalation, but rather that the weaker party escalates against the stronger adversary to provoke an over-reaction that can spur support for the weaker party's agenda, at the same time that it triggers the stronger adversary into a costly war that it had preferred to avoid. There are two versions of this rationale.

First, the weaker alliance member that is threatened by a stronger adversary needs to escalate a potential crisis in order to avoid being abandoned if the allies think that the level of conflict is too low to intervene. This was, as Rolf Tamnes recognized, very much the situation of Cold War Norway. If the Soviet Union had been able to keep a potential conflict to a low level of violence, Norway was in danger of having to face the Soviets alone. An isolated conflict would have resulted in an overwhelming superiority for the Soviet Union where Norway would have been outnumbered in virtually all domains. Hence, in order to drag the rest of NATO into a conflict, Norway would have had to escalate the conflict to provoke an over-reaction.[32] Weak party escalation is thus rational and strategic if the escalation in and of itself creates a situation in which the weaker party receives outside help.

Second and related, it is rational for the weaker alliance member to escalate to provoke an over-reaction that will force the stronger adversary to over-stretch. Rather than building upon a fear of being left alone, this logic of provocation builds upon the weaker party escalating to provoke an over-reaction that will leave the adversary to split its forces and thus be weaker on several fronts. Again, the logic rests upon the degree to which the weaker party can rely upon other alliance members to join in the fighting. It is rational for a hyena to challenge the lion for its prey if it gets help from the rest of the pack and several hyenas can have a bite, while the lion chases the most obvious challenger. If the hyenas can take turns and

---

[31]Thomas Rid, *Cyber War Will Not Take Place* (Oxford: Oxford University Press 2013); P.W. Singer and Allan Friedman, *Cybersecurity and Cyberwar: What Everyone Needs to Know* (Oxford: Oxford University Press 2014); Thomas Rid and Ben Buchanan, 'Attributing Cyber Attacks', *Journal of Strategic Studies* 39/1 (2015), 4–37.

[32]Tamnes, *The United States and the Cold War*.

escalate against the lion, it will over-stretch and the hyenas collectively can have a feast on the lion's behalf.

In empirical terms, this third logic can be witnessed in the wars of national liberation in the 1960s and perhaps most visibly in the Portuguese colonies.[33] Each of Angola, Mozambique, Guinea-Bissau and Cape Verde were individually weaker than the Colonial power Portugal. However, by escalating against Portugal in the same period of time, it became untenable for the Portuguese army to maintain its colonies. Guinea-Bissau was first to rebel and the war raged between 1956 and 1974, Angola fought its war of independence between 1961 and 1974, and Mozambique 1964–1974.[34]

Hence, as long as the colonial power fell for the provocation and waged war in several directions, it became rational for the next Colony to pick up weapons and escalate. Note that this logic differs from the case of Finland's escalation against the Soviet Union since Finland had aligned interests with Germany that were on equal footing, and in some respects superior, with the Soviet Army. Neither of the former Portuguese Colonies were in such a position.

The strategic logic of building reputation

The final strategic logic for weak parties to escalate against stronger adversaries relates to the importance of building a long-term reputation of being steadfast and belligerent. It is rational to escalate if the short-term costs incurred are lower than the long-term benefits. In situations where the weak party does not have the benefits of alliances or where it may not be possible to compartmentalize conflicts, it can still be rational to escalate against strong adversaries. The reason is that the long-term costs of yielding in a militarized crisis can be long-term bullying by the stronger party. Bullying in world politics can, of course, be understood as indefinite imposed costs. If these costs are higher than the costs of escalation into a potentially even unwinnable short-term war, then it is rational to escalate.

Reputation, research suggests, operates along two related, but not exactly similar causal mechanisms towards escalation. First, as suggested above, in order to prevent a much worse long-term future, it is rational for the weak party to escalate short-term if it can build long-term credibility of being belligerent. Such a reputation can improve the likelihood of the weak party successfully deterring the stronger in future interactions.

Consider, again, the North Korean refusal to abide with anything resembling international norms of non-spread of nuclear weapons. It is rational for North Korea to escalate against the United States in the form of verbal

---

[33] Al J. Venter, *Portugal's Guerilla Wars in Africa* (London: Helion, 2013).
[34] John P. Cann, *Counterinsurgency in Africa: The Portuguese way of war, 1961–1974* (London: Greenwood Press, 1997).

threats, new nuclear tests and ballistic missile testing if its credibility to deter a long-term defeat and loss of independence is increased by escalation. In this case, escalation increases the credibility of actually launching a missile over Japan or South Korea that can carry nuclear warheads, which in and of itself is crucial for its nuclear deterrence. Sabre rattling in the face of supreme power is thus rational. Even if North Korea slowly recourse from its threatening posture and stops overt missile testing, its long-term reputation of not backing down, may reap long-term deterrent advantages.

Second, as several scholars have pointed out, maintaining a particular reputation is also a motive for escalating to war.[35] In this case, it is not escalation to create a reputation of not yielding, but rather weak party escalation to maintain a reputation and thus a long-term credibility to deter future bullying. Most importantly, it is by having the reputation of being steadfast and credible that you can maintain a long-term reputation even if you muster short-term defeat.

One, perhaps less well-known, expression of this logic is the War of the Triple Alliance in which Paraguay simultaneously fought Brazil, Uruguay and Argentina in 1864–1870. Fearing a rapprochement of the region's two great powers Brazil and Argentina, Paraguay faced a long-term future with a growing negative power balance, which very well may have included being divided between the great powers. As a result, it launched an attack while still being in comparison relatively strong, yet in nominal terms still much weaker. As Weiziger shows, many contemporary explanations of then President Lopez's decision as simply crazy are misleading and even if he held ideas of being successful in the short-term, his greatest concern was the long-term survival of the country.[36] As such, escalation to forge a reputation of being belligerent was preferable to be perceived as weak and suffer hardships under a future peace.

Weak-party escalation: Rational and strategic, but rare

In this section, we will elaborate on a series of dilemmas of the weak stemming from the four logics of weak party escalation. It is also these dilemmas that cause weak party escalation to be rare, despite the fact that it can be rational and, thus, strategic.

---

[35]Allan Dafoe, Jonathan Renshon, and Paul Huth, 'Reputation and Status as Motives for War', *Annual Review of Political Science* 17 (2014), 371–93; Allan Dafoe and Devin Caughey, 'Honor and War: Southern US Presidents and the Effects of Concern for Reputation', *World Politics* 68/2 (2016), 341–81.
[36]Alex Weisiger, *Logics of War: Explanations for Limited and Unlimited Conflicts* (Ithaca, NY: Cornell University Press 2013), 86–104.

Credible commitment issues

Several of the dangers involved in weak party escalation are related to so-called 'credible commitment problems'. Not being able to trust other states' long-term intentions due to inherent problems of credibly signalling commitments for cooperation, Fearon argued, led states to launch preventive wars since one feared a future unfavourable peace.[37] Commitment problems, according to rationalist theory, can be avoided by so-called 'costly signalling', such as self-imposed punishments in case of defection or agreements where remuneration is specified for defecting from agreements. Such agreements are commonly part of alliances and alliance membership can be understood as attempts to avoid problems related to credible commitments.[38]

It is precisely problems of credible commitment from fellow alliance members or otherwise friendly and sympathetic great powers that make weak party escalation a rare phenomenon. First, both opportunist escalation and provocation escalation relies upon outside assistance. The likelihood of receiving such support, however, diminishes if the weak party is perceived as the aggressor in a dispute; in classical alliance theory a risk of 'entrapment' that the allied major power will avoid.[39]

To be stood up by one's 'big brother' – rightly or wrongly – could be disastrous for the weak party in an asymmetric conflict. It would mean that the weak party rolls the dice and escalates and no one is there to back it up once a conflict starts. The inherent dangers of escalation under such circumstances ought to be clear for everyone and problems of credible commitment, therefore, inhibit weak party escalation, making it rare.

Credible commitment issues are also at the heart of another fear of the would-be small state escalator. If the aim is to compartmentalize escalation into one domain where one possesses escalation dominance, the ability to deter cross-domain escalation on behalf of the strong party will be critical. The weak party will potentially be worse off if it escalates in one domain to solve problems inherent and possibly limited to that domain, if the strong party as a response escalates in other domains. The ability to compartmentalize issues is therefore crucial, but there are few issues related to strategy and security, where a stronger party is willing to abide by rules and limits set by the weak. The ability to deter cross-domain escalation on behalf of the stronger opponent

---

[37]James D. Fearon, 'Rationalist Explanations for War', *International Organization* 49/3 (1995), 379–414.
[38]Glenn Snyder, *Alliance Politics* (Ithaca, NY: Cornell University Press 2007).
[39]Glenn Snyder, 'The security dilemma in alliance politics', *World Politics*, 36/4 (1984), 461–95.

relies as all deterrence on credibility and communication,[40] and weak parties, in particular, usually struggle to deter stronger adversaries successfully.

The obvious solution for improving deterrence for many small states have traditionally been to group together in alliances to improve the credibility of deterrence and it is in this way, the absence of credible commitments also becomes a problem for those weak parties that try escalate within one domain. As a result, few small states will be prone to take risks under such conditions.

Information shortages and assessment errors

Even if we assume that credible commitment problems did not exist and escalation is a rational, interest-driven decision and that actors occasionally strive to reap long-term dividends on the expense of short-term costs, it does not follow that actors get it right all the time. On the contrary, information, or, more precisely, lack of information, seems endemic in world affairs and such problems, of course, create flawed estimates of short-term costs and long-term benefits. States 'operate in a world of imperfect information, where potential adversaries have incentives to misrepresent their own strength or weakness and to conceal their true aims.'[41]

Incentives to manipulate information may be even harder to resist when stakes are high, for example, in security issues, since the possible gains of bluffing become higher when the stakes are raised. Such so-called 'information failures' are thus inherent to strategic issues since it is in states' interests to maintain private information about, for example, the exact capabilities of their armed forces and indeed about their future intentions.[42] Lying and cheating can be prosperous in strategy.

The greatest fear for those pursuing long-term gains at the costs of short-term costs is that the short-term costs turn out to be higher than expected and the long-term gains lower than expected. The realization that all states have incentives to manipulate information also have a conservative consequence. States are less prone to take risks if they know that the counterparts always have some aces up their sleeves. The consequences of being wrong for a small state, in particular, can be grave and therefore, the same dynamics that occasionally make it rational

---

[40]See, e.g., George and Smoke, *Deterrence in American Foreign Policy*; Lawrence Freedman, *Deterrence* (Cambridge: Polity Press 2004); Patrick Morgan 'The State of Deterrence in International Politics Today', *Contemporary Security Policy* 33/1 (2012), 85–107.

[41]John J. Mearsheimer, 'The False Promise of International Institutions', *International Security* 19/3 (1995), 10.

[42]James D. Fearon, 'Rationalist Explanations for War'.

to escalate also make it rational not to escalate against the strong. Weak party escalation is thus rational, but rare.

Conclusions

Why do weak parties escalate against stronger adversaries? In this article, we have outlined the strategic logic of four rationalist explanations of weak party escalation: provocation, compartmentalization, the division of labour, and reputation. We have also elaborated on the dilemmas or dangers of the weak stemming from the four logics of weak party escalation: being perceived as the aggressor, limited ability to control war, and limited ability to calculate short-term costs. By exploring the logic of weak party escalation as well as the dangers of escalation, we have also identified some of the scope conditions for weak party escalation.

The main result of the article is that weak parties' strategic options are wider than commonly thought. Weak parties, especially small states, have often been described as 'passive victims' in international affairs, totally in the hands of great power competition and power play. As Thucydides put it: 'The strong do what they can, and the weak suffer what they must.' We argue that this view should be nuanced and that leaders of small states should realize that resistance and escalation, even against great powers, can be beneficial – even if the final result is short-term defeat.

The question is, however, if political leaders are prepared to escalate and if generals are prepared to advice escalation. In a dictatorship, such as North Korea, this might be the case, but in a democracy, it may be less likely. However, we argue that it is important for policymakers in democratic weaker party relations as well, to realize that escalation can be a rational strategy amongst others in a conflict. This is not a common view by decision makers in small states today, but our point is that it will increase the strategic options for such policymakers to achieve their strategic goals. That is not to say, of course, that small states should escalate conflicts but just to inform decision-makers in small states that there are more options than just be a passive victim.

More research should be conducted within the field of small state strategy, rationality and escalation, not only because research on small state strategies is rare compared to research on great power strategies, but because there is a need to show practitioners – political and military leaders – that they can better achieve their goals if they think and act strategically. In particular, in order to properly test the theorizing in this article systematically, there is a great need for comparative case studies. We stress this research design in favour of quantitative studies for the simple reason that we suspect that weak party escalation is too rare so comparative case studies are more advantageous. We should also stress that in order to correctly assess strategic elites' decisions to escalate, it is probably also necessary with a combination of archival research for older cases and elite inter-views for more current ones.

Future research also needs to take into consideration if the way that escalation is carried out matters insofar as achieving various forms of strategic effects. It should also be recognized that the different causal trajectories explored and developed here only seek to understand weak party escalation as a result of external dynamics. There should also be efforts to uncover if and how domestic politics can be a cause of escalation. Hence, although we have contributed by analyzing and discussing weak party escalation, much effort remains.

Disclosure statement

No potential conflict of interest was reported by the authors.

Bibliography

Angstrom, Jan, 'Evaluating Rivalling Interpretations of Asymmetric War and Warfare', Karl-Erik Haug and Ole Jorgen Maao (eds.), *Conceptualising Modern War*, 29–48 (New York: Columbia University Press 2011).
Arreguin-Toft, Ivan, 'How the Weak Win Wars: A Theory of Asymmetric Conflict', *International Security* 26/1 (2001), 93–128. doi: 10.1162/016228801753212868.
Arreguin-Toft, Ivan, *How the Weak Win Wars: A Theory of Asymmetric Conflict* (Cambridge: Cambridge University Press 2005).
Betts, Richard K., 'Should Strategic Studies Survive?', *World Politics* 50/1 (1997), 7–33. doi: 10.1017/S0043887100014702.
Blainey, Geoffrey, *The Causes of War*, 3rd (New York: The Free Press 1988).
Børresen, Jacob, *Torskekrig!: Om forutsetninger og rammer for kyststatens bruk av makt* (Oslo: Abstrakt 2011).
Brodie, Bernard, *Escalation and the Nuclear Option* (Princeton, NJ: Princeton University Press 1966).
Cann, John P., *Counterinsurgency in Africa: The Portuguese Way of War, 1961–1974* (London: Greenwood Press 1997).

Dafoe, Allan and Devin Caughey, 'Honor and War: Southern US Presidents and the Effects of Concern for Reputation', *World Politics* 68/2 (2016), 341–81. doi: 10.1017/S0043887115000416.

Dafoe, Allan, Jonathan Renshon, and Paul Huth, 'Reputation and Status as Motives for War', *Annual Review of Political Science* 17 (2014), 371–93. doi: 10.1146/annurev-polisci -071112-213421.

Davis, Carmel, 'An Introduction to Nuclear Strategy and Small Nuclear Powers: Using North Korea as a Case', *Defence Studies* 9/1 (2009), 93–117. doi: 10.1080/14702430802666694.

Duyvesteyn, Isabelle, 'The Escalation and De-Escalation of Irregular War: Setting Out the Problem', *Journal of Strategic Studies* 35/5 (2012), 601–12. doi: 10.1080/01402390.2012.706750.

Fearon, James D., 'Rationalist Explanations for War', *International Organization* 49/3 (1995), 379–414. doi: 10.1017/S0020818300033324.

Fischerkeller, Michael P., 'David versus Goliath: Cultural Judgements in Asymmetric Wars', *Security Studies* 7/4 (1998), 1–43. doi: 10.1080/09636419808429357.

Freedman, Lawrence, 'On the Tiger's Back: The Development of the Concept of Escalation', Roman Kolkowicz (ed.), *The Logic of Nuclear Terror* (London: Allen & Unwin 1987), 109–52.

Freedman, Lawrence, *Deterrence* (Cambridge: Polity Press 2004).

Fuller, J.F.C., *The Foundaitions of the Science of War* (London: Hutchinson 1926).

George, Alexander and Richard Smoke, *Deterrence in American Foreign Policy* (New York: Columbia University Press 1974).

Gow, James, *The Serbian Project and Its Adversaries: A Strategy of War Crimes* (London: Hurst 2003).

Harrison, Wagner, R., 'Bargaining and War', *American Journal of Political Science* 44/3 (2000), 469–84.

Hatton, Ragnhild Marie, *Charles XII of Sweden* (London: Weidenfield and Nicolson 1968).

Herz, John, 'Idealist Internationalism and the Security Dilemma', *World Politics* 2/2 (1950), 171–201.

Kahn, Herman, *On Thermonuclear War* (Princeton, NJ: Princeton University Press 1960).

Kahn, Herman, *On Escalation: Metaphors and Scenarios* (New York: Praeger 1965).

Kaplan, Fred, *The Wizards of Armageddon* (Stanford, CA: Stanford University Press 1983).

Kaplan, Morton, *The Strategy of Limited Retaliation* (Princeton, NJ: Woodrow Wilson School 1959).

Kilgour, Mark and Frank Zagare, 'Explaining Limited Conflicts', *Conflict Management and Peace Science* 24/1 (2007), 65–82. doi: 10.1080/07388940601102852.

Knight, Roger, *Britain against Napoleon: The Organization of Victory* (London: Penguin 2014).

Kühn, Ulrich, *Preventing Escalation in the Baltics: A NATO Playbook* (Washington, DC: Carnegie 2018).

Lake, David and Donald Rothchild eds., *The International Spread of Ethnic Conflict: Fear, Diffusion, and Escalation* (Princeton, NJ: Princeton University Press 1998).

Mack, Andrew, 'Why Big Nations Lose Small Wars: The Politics of Asymmetric Conflict', *World Politics* 27/2 (1975), 175–200. doi: 10.2307/2009880.

Mearsheimer, John, *The Tragedy of Great Power Politics* (New York: Norton 2001).

Mearsheimer, John J., 'The False Promise of International Institutions', *International Security* 19/3 (1995), 5–49.

Meinander, Carl Henrik, 'Finland', David Stahel (ed.), *Joining Hitler's Crusade: European Nations and the Invasion of the Soviet Union, 1941*, 17–45 (Cambridge: Cambridge University Press 2017).

Morgan, Patrick, 'The State of Deterrence in International Politics Today', *Contemporary Security Policy* 33/1 (2012), 85–107. doi: 10.1080/13523260.2012.659589.

Morgenthau, Hans, *Politics among Nations: The Struggle for Power and Peace* (New York: McGraw-Hill 1948).

Paul, T.V., *Asymmetric Conflicts: War Initiation by Weaker Powers* (Cambridge: Cambridge University Press 1994).

Powell, Robert, 'Bargaining Theory and International Conflict', *Annual Review of Political Science* 5 (2002), 1–30. doi: 10.1146/annurev.polisci.5.092601.141138.

Reiter, Dan, 'Exploring the Bargaining Model of War', *Perspectives on Politics* 1/1 (2003), 27–43. doi: 10.1017/S1537592703000033.

Rid, Thomas, *Cyber War Will Not Take Place* (Oxford: Oxford University Press 2013).

Rid, Thomas and Ben Buchanan, 'Attributing Cyber Attacks', *Journal of Strategic Studies* 39/1 (2015), 4–37. doi: 10.1080/01402390.2014.977382.

Schelling, Thomas, *Arms and Influence* (New Haven, CT: Yale University Press 1966).

Schweller, Randall L., 'Bandwagoning for Profit: Bringing the Revisionist State Back', *International Security* 19/1 (1994), 72–107. doi: 10.2307/2539149.

Singer, P.W. and Allan Friedman, *Cybersecurity and Cyberwar: What Everyone Needs to Know* (Oxford: Oxford University Press 2014).

Slantchev, Branislav, *Military Threats: The Costs of Coercion and Price for Peace* (Cambridge: Cambridge University Press 2011).

Smith, Mike, 'Escalation in Irregular War: Using Strategic Theory to Examine from First Principles', *Journal of Strategic Studies* 35/5 (2012), 613–38. doi: 10.1080/01402390.2012.706967.

Smith, Rupert, *The Utility of Force: The Art of War in the Modern World* (London: Allen Lane 2005).

Smoke, Richard, *War: Controlling Escalation* (Cambridge, MA: Harvard University Press 1977).

Snyder, Glenn, 'The Security Dilemma in Alliance Politics', *World Politics* 36/4 (1984), 461–95. doi: 10.2307/2010183.

Snyder, Glenn, *Alliance Politics* (Ithaca, NY: Cornell University Press 2007).

Szabo, Franz A.J., *The Seven-Years War in Europe, 1756–1763* (London: Routledge 2008).

Tamnes, Rolf, *The United States and the Cold War in the High North* (Aldershot: Palgrave 1991).

von Clausewitz, Carl, *On War* translated by Michael Howard and Peter Paret (Princeton, NJ: Princeton University Press 1976).

Waltz, Kenneth, *Theory of International Politics* (New York: McGraw-Hill 1979).

Weisiger, Alex, *Logics of War: Explanations for Limited and Unlimited Conflicts* (Ithaca, NY: Cornell University Press 2013).

Wilkinson, Benedict and James Gow eds., *The Art of Creating Power: Freedman on Strategy* (London: Hurst 2017).

Index

For Product Safety Concerns and Information please contact our EU
representative GPSR@taylorandfrancis.com
Taylor & Francis Verlag GmbH, Kaufingerstraße 24, 80331 München, Germany

www.ingramcontent.com/pod-product-compliance
Lightning Source LLC
Chambersburg PA
CBHW060314220326
41598CB00027B/4327